Macmillan/McGraw-Hill • Glencoe

Grade
4

Math Triumphs

Book 3: Measurement

Authors

Basich Whitney • Brown • Dawson • Gonsalves • Silbey • Vielhaber

**Macmillan/McGraw-Hill
Glencoe**

Photo Credits

All coins photographed by United States Mint.
All bills photographed by Michael Houghton/StudiOhio.
Cover, i BananaStock/Jupiter Images; **iv** (tl)File Photo, (tc tr)The McGraw-Hill Companies, (cl c)Doug Martin, (cr)Aaron Haupt, (bl bc)File Photo; **v** (L to R 1 2 3 4 6 7 8 9 11 12)The McGraw-Hill Companies, (5 10 13 14)File Photo; **vi** Index Stock/Alamy; **vii** Digital Vision; **viii** Brand X Pictures/Alamy; **308** Nancy R. Cohen/Getty Images; **309** Photodisc/Getty Images; **314** Ingram Publishing/SuperStock; **318** CORBIS; **319** (t)Larry Mulvehill/CORBIS, (b)George Gutenberg/Beateworks/CORBIS; **324** Photodisc/Getty Images; **325** (t)Ryan McVay/Getty Images, (b)Photos.com/Jupiter Images; **326** (tr)Getty Images, (tc)Stockbyte, (br)Scot Frei/CORBIS; **330** John A. Rizzo/Getty Images; **334** Stockbyte; **339** Tim Pannell/CORBIS; **350** redcover.com/Getty Images; **356** Hisham F. Ibrahim/Getty Images; **358** (t)Kokyat Choong/The Image Works, (b)Martinique & Guadeloupe/Iconotec; **364** Leland Bobbe/Photonica/Getty Images; **370** Michael Newman/PhotoEdit; **372** Juniors Bildarchiv/Alamy; **380** Joel Sartore/National Geographic/Getty Images; **381** Stockbyte; **388** C Squared Studios/Getty Images; **394** Apis/Abramis/Alamy; **396** Thinkstock/CORBIS; **401** Creatas/PunchStock; **407** Michael Newman/PhotoEdit; **409** Burke/Triolo Productions/Brand X Pictures/Getty Images; **410** Siede Preis/Getty Images

The McGraw-Hill Companies

 **Macmillan/McGraw-Hill
Glencoe**

Send all inquiries to:
Glencoe/McGraw-Hill
8787 Orion Place
Columbus, OH 43240-4027

ISBN: 978-0-07-888203-6
MHID: 0-07-888203-6

Printed in the United States of America.

Math Triumphs
Grade 4, Book 3

2 3 4 5 6 7 8 9 10 066 16 15 14 13 12 11 10 09 08

Book 1

Book 2

Book 3

Authors and Consultants

AUTHORS

Frances Basich Whitney
Project Director, Mathematics K–12
Santa Cruz County Office of Education
Capitola, California

Kathleen M. Brown
Math Curriculum Staff Developer
Washington Middle School
Long Beach, California

Dixie Dawson
Math Curriculum Leader
Long Beach Unified
Long Beach, California

Philip Gonsalves
Mathematics Coordinator
Alameda County Office of Education
Hayward, California

Robyn Silbey
Math Specialist
Montgomery County Public Schools
Gaithersburg, Maryland

Kathy Vielhaber
Mathematics Consultant
St. Louis, Missouri

CONTRIBUTING AUTHORS

Viken Hovsepian
Professor of Mathematics
Rio Hondo College
Whittier, California

FOLDABLES
Study Organizer
Dinah Zike
Educational Consultant
Dinah-Might Activities, Inc.
San Antonio, Texas

CONSULTANTS

Assessment

Donna M. Kopenski, Ed.D.
Math Coordinator K–5
City Heights Educational Collaborative
San Diego, California

Instructional Planning and Support

Beatrice Luchin
Mathematics Consultant
League City, Texas

ELL Support and Vocabulary

ReLeah Cossett Lent
Author/Educational Consultant
Alford, Florida

Reviewers

Each person below reviewed at least two chapters of the Student Study Guide, providing feedback and suggestions for improving the effectiveness of the mathematics instruction.

Dana M. Addis
Teacher Leader
Dearborn Public Schools
Dearborn, MI

Renee M. Blanchard
Elementary Math Facilitator
Erie School District
Erie, PA

Jeanette Collins Cantrell
5th and 6th Grade Math Teacher
W.R. Castle Memorial Elementary
Wittensville, KY

Helen L. Cheek
K-5 Mathematics Specialist
Durham Public Schools
Durham, NC

Mercy Cosper
1st Grade Teacher
Pershing Park Elementary
Killeen, TX

Bonnie H. Ennis
Mathematics Coordinator
Wicomico County Public Schools
Salisbury, MD

Sheila A. Evans
Instructional Support Teacher—Math
Glenmount Elementary/Middle School
Baltimore, MD

Lisa B. Golub
Curriculum Resource Teacher
Millennia Elementary
Orlando, FL

Donna Hagan
Program Specialist—Special Programs
   Department
Weatherford ISD
Weatherford, TX

Russell Hinson
Teacher
Belleview Elementary
Rock Hill, SC

Tania Shepherd Holbrook
Teacher
Central Elementary School
Paintsville, KY

Stephanie J. Howard
3rd Grade Teacher
Preston Smith Elementary
Lubbock, TX

Rhonda T. Inskeep
Math Support Teacher
Stevens Forest Elementary School
Columbia, MD

Albert Gregory Knights
Teacher/4th Grade/Math Lead Teacher
Cornelius Elementary
Houston, TX

Barbara Langley
Math/Science Coach
Poinciana Elementary School
Kissimmee, FL

David Ennis McBroom
Math/Science Facilitator
John Motley Morehead Elementary
Charlotte, NC

Jan Mercer, MA; NBCT
K-5 Math Lab Facilitator
Meadow Woods Elementary
Orlando, FL

Rosalind R. Mohamed
Instructional Support Teacher—Mathematics
Furley Elementary School
Baltimore, MD

Patricia Penafiel
Teacher
Phyllis Miller Elementary
Miami, FL

Lindsey R. Petlak
2nd Grade Instructor
Prairieview Elementary School
Hainesville, IL

Lana A. Prichard
District Math Resource Teacher K-8
Lawrence Co. School District
Louisa, KY

Stacy L. Riggle
3rd Grade Spanish Magnet Teacher
Phillips Elementary
Pittsburgh, PA

Wendy Scheleur
5th Grade Teacher
Piney Orchard Elementary
Odenton, MD

Stacey L. Shapiro
Teacher
Zilker Elementary
Austin, TX

Kim Wilkerson Smith
4th Grade Teacher
Casey Elementary School
Austin, TX

Wyolonda M. Smith, NBCT
4th Grade Teacher
Pilot Elementary School
Greensboro, NC

Kristen M. Stone
3rd Grade Teacher
Tanglewood Elementary
Lumberton, NC

Jamie M. Williams
Math Specialist
New York Mills Union Free School District
New York Mills, NY

Contents

Chapter 8 — Geometry and Measurement

Ashepoo River, Walterboro, South Carolina

Chapter 9 · Area

Badlands National Park, Interior, South Dakota

Contents

Chapter 10 — **Spatial Reasoning**

Niagara Falls, Niagara Falls, New York

SCAVENGER HUNT

CHAPTER 1

Let's Get Started

Use the Scavenger Hunt below to learn where things are located in each chapter.

1. What is the title of Chapter 8?

2. What is the Key Concept of Lesson 10-3?

3. What is the definition of area on page 366?

4. What are the vocabulary words for Lesson 9-2?

5. How many examples are presented in the Chapter 10 Study Guide?

6. What item is being measured in Example 2 on page 328?

7. What shapes are being used to make a hexagon in Example 1 on page 352?

8. What do you think is the purpose of the Progress Check 2 on page 410?

9. On what pages will you find the study guide for Chapter 9?

10. In Chapter 8, find the internet address that tells you where you can take the Online Readiness Quiz.

Geometry and Measurement

Lines, angles, and shapes surround us.

We can measure lines, angles, and shapes. These measurements help us build fences, homes, bicycles, and playground equipment we use every day.

STEP **1** Quiz

Are you ready for Chapter 8? Take the Online Readiness Quiz at *macmillanmh.com* to find out.

STEP **2** Preview

Get ready for Chapter 8. Review these skills and compare them with what you'll learn in this chapter.

What You Know	What You Will Learn
You know how to describe and recognize some figures. **TRY IT!** Describe each figure. **1** _____ _____ **2** _____ _____	*Lessons 8-1 and 8-3* **Polygons** are closed figures. They have sides made of **line segments**. A **rectangle** is made of 4 line segments. 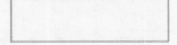
You know how to use a number line. **Example:** 2 + 3 = 5 0 1 2 3 4 5 6 7 8 9 **TRY IT!** Use the number line to add and subtract. 0 1 2 3 4 5 6 7 8 9 **3** $4 + 5 =$ _____ **4** $8 - 4 =$ _____	*Lesson 8-4* You will use the "number line" on a **ruler** to measure objects. 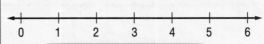 0 1 2 3 4 5 6 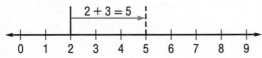 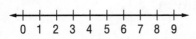 0 1 2 3 4 5 6 7 cm The eraser is 5 centimeters long.

Lines

KEY Concept

The vocabulary words below describe the types of figures and the relationships between lines.

Vocabulary Word	Example	Information
point	Z • point Z	A **point** is named using capital letters. It is represented by dots.
line	A ←——→ B line AB	A **line** can be named by two of the points on the line.
line segment	A •——• B line segment AB	A **line segment** is named by the points on the ends of the line.
parallel lines	←——→ ←——→	**Parallel lines** do not cross.
intersecting lines	✕ C	**Intersecting lines** cross at a point. These lines cross at point C.

VOCABULARY

intersecting lines
　lines that meet or cross each other

line
　a set of points that goes straight in opposite directions without ending

line segment
　a part of a line between two endpoints

parallel lines
　lines that are the same distance apart; parallel lines do not meet or cross

point
　an exact location in space

Example 1

Draw a line. Name the figure line QR.

1. Use a ruler to draw a line.
2. Draw arrows on each end of the line to show that the line does not end in either direction.
3. Draw two points on the line. Label the points Q and R.

　　　Q　　　　R
　←——•————•——→

YOUR TURN!

Draw a line segment. Name the figure line segment ST.

1. Use a ruler to draw a line.
2. Draw _____ of the line to show that the line _____.
3. Draw _____ on the line. Label the points _____ and _____.

Example 2

Identify the relationship.

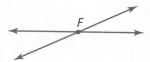

1. The two lines cross at point *F*.
2. Lines that cross are called intersecting lines.

YOUR TURN!

Identify the relationship.

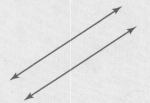

1. The two lines _____.
2. Lines that _____ are called _____ lines.

Who is Correct?

Identify the figure.

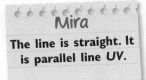

Mira

The line is straight. It is parallel line *UV*.

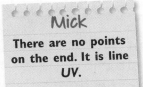

Mick

There are no points on the end. It is line *UV*.

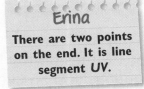

Erina

There are two points on the end. It is line segment *UV*.

Circle correct answer(s). Cross out incorrect answer(s).

 Guided Practice

Draw each figure.

1 intersecting lines

2 line *JK*

GO ON

Step by Step Practice

3 **Identify the relationship.**

Step 1 The two lines _____ cross.

Step 2 Lines that _____ are called _____ lines.

Identify the relationship of each set of two lines.

4 _____

5 _____

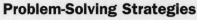

Step by Step Problem-Solving Practice

Solve.

6 **MAPS** Pia lives on Maple Street. Her aunt lives on Elm Street. What type of lines do the two streets form? Use the map to help you answer the question.

Problem-Solving Strategies

✓ Use a diagram.
☐ Look for a pattern.
☐ Act it out.
☐ Solve a simpler problem.
☐ Work backward.

Understand   Read the problem. Write what you know.

The streets in the exercise are _____ and _____.

Plan   Pick a strategy. One strategy is to use a diagram.

Solve   Look at the diagram. Do the streets meet or intersect? _____

So, the streets form _____ lines.

Check   Use your real-world knowledge. What word names the place at which two streets meet?

_____

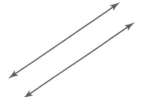

7 **MAPS** Keisha walked her dog from her house to the park. Use a ruler to draw a line for Keisha's path. Ursula walked from her house to the library. Use a ruler to draw a line for Ursula's path. What type of lines do the two paths form? Check off each step.

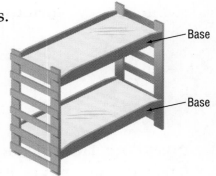

_____ Understand: I circled key words.

_____ Plan: To solve the problem, I will _____.

_____ Solve: The answer is _____.

_____ Check: I will check my answer by _____.

8 **DESIGN** Mr. Inez bought bunk beds for his twin sons. What kind of lines are the two bases in the picture at the right?

_____

9 **Reflect** Are the two lines shown below intersecting lines? Explain.

_____

_____

▶ Skills, Concepts, and Problem Solving

Draw each figure.

10 parallel lines

11 line segment *LM*

12 point C

13 line AB

GO ON

Identify the relationship of each set of lines.

14 _____

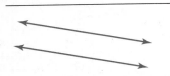

15 _____

Solve.

16 **HOBBIES** Kelsey's Kite Shop designs kites.
What is the relationship between the lines on the kite?

_____

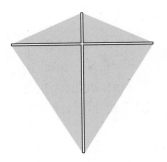

17 **CONSTRUCTION** Doug has an iron
fence that surrounds his house and
yard. What type of lines do the
vertical iron bars form?

_____

Vocabulary Check **Write the vocabulary word that completes
each sentence.**

18 A(n) _____ is an exact location in space.

19 _____ lines are lines that cross or meet at a point.

20 **Writing in Math** Explain how to draw line segment *JK*.

_____

_____

STOP

KEY Concept

Angles are created when two lines or two line segments meet or cross. They can be grouped by their measures.

Vocabulary Word	Example	Information
right angle	This symbol indicates a right angle.	A right angle measures exactly 90°. The sharp corner of your book forms a right angle.
acute angle		An acute angle measures between 0° and 90°. It is an angle smaller than a right angle.
obtuse angle		An obtuse angle measures between 90° and 180°. It is an angle greater than a right angle.
straight angle		A straight angle measures exactly 180°. It forms a line.

VOCABULARY

angle
a figure formed when two lines or line segments meet or cross

degree (°)
a unit for measuring angles
Example:
90 degrees = 90°

vertex
the common endpoint of the sides of an angle (the plural is *vertices*)

Example 1

Identify the type of angle.

1. Does the angle have a square corner? No. So, it is not a right angle.

2. The angle looks greater than 90 degrees. It is a(n) obtuse angle.

YOUR TURN!

Identify the type of angle.

1. Does the angle look like a right angle? _____

2. The angle looks _____ 90 degrees. It is a(n) _____ angle.

GO ON

Example 2

Use grid paper to draw an acute angle.

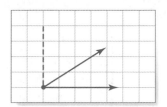

1. Draw one line along a line on the grid.

2. Draw a dashed line to create a right angle. To draw an acute angle, draw the second line between the lines.

3. Does your angle look smaller than a right angle? Yes; it is acute.

YOUR TURN!

Use grid paper to draw an obtuse angle.

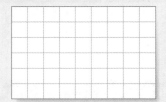

1. Draw one line _____ on the grid.

2. Draw a dashed line to create a right angle. To draw an _____ angle, draw the second line outside the lines.

3. Does your angle look _____ than a right angle?

_____

Who is Correct?

Identify the type of angle.

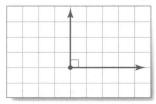

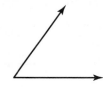

Rachel
It is a straight angle.

Ivan
It is a right angle.

Deon
It is an obtuse angle.

Circle correct answer(s). Cross out incorrect answer(s).

Guided Practice

Identify each type of angle.

1 _____

2 _____

Step by Step Practice

3 **Use grid paper to draw a straight angle.**

Step 1 Start at a point. Draw a line to the right along a line on the grid.

Step 2 Draw a line to the left along the same line on the grid.

Step 3 Does the straight angle look like a line? Yes, it is a straight angle.

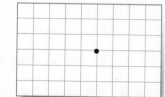

Use grid paper to draw each angle.

4 right angle

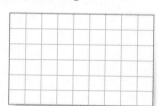

5 acute angle

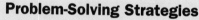

Step by Step Problem-Solving Practice

Solve.

Problem-Solving Strategies
✓ Use a diagram.
☐ Look for a pattern.
☐ Guess and check.
☐ Solve a simpler problem.
☐ Work backward.

6 **BASEBALL** Quincy plays baseball. He noticed that the lines from first to second base and second to third base form an angle. What type of angle is formed at second base?

Understand Read the problem. Write what you know.

Quincy was looking at the angle formed by the line from _____ base to _____ base and from _____ base to _____ base.

Plan Pick a strategy. One strategy is to use a diagram.

Solve Look at the diagram. The angle formed at second base is a(n) _____ angle.

Check Review the definition of a _____ angle.

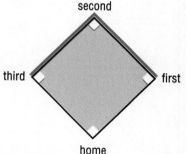

GO ON

7 BILLIARDS How many acute angles does a pool rack have? Check off each step.

_____ Understand: I circled key words.

_____ Plan: To solve the problem, I will _____.

_____ Solve: The answer is _____.

_____ Check: I will check my answer by _____.

8 TIME Justin eats lunch at school at 11:15. What type of angle do the hands on a clock form when the time is 11:15?

_____

9 Reflect Ruben says that the measure of this angle is 105 degrees. Explain Ruben's error.

_____

_____

_____

▶ Skills, Concepts, and Problem Solving

Identify each type of angle.

10 _____

11 _____

12 _____

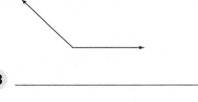

13 _____

Use grid paper to draw each angle.

14 straight angle

15 obtuse angle

Solve.

16 SAILING   What types of angles are formed by the sides of the sails shown at the right?

_____

17 CARS   After back surgery, Kono could not sit up straight while riding in a car. He had to recline the seat at an angle that measured 110 degrees. What type of angle was Kono's reclining position?

_____

SAILING This sailboat has triangular sails.

Vocabulary Check   Write the vocabulary word that completes each sentence.

18 A(n) _____ is formed when two lines or two line segments meet or cross.

19 A(n) _____ angle is an angle that measures greater than 90 degrees but less than 180 degrees.

20 A(n) _____ angle is an angle that measures greater than 0 degrees but less than 90 degrees.

▶ Spiral Review

Solve. (Lesson 8-1, p. 310)

21 FENCES   Derek's horse runs in a field that is surrounded by a fence. What type of lines do the slats on the fence form? _____

Identify the relationship of the set of lines.

22 _____

FENCES Derek's horse runs in the field surrounded by this fence.

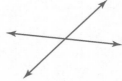

STOP

Draw each figure.

1 line *AB*

2 line segment *FG*

Use grid paper to draw each angle.

3 obtuse

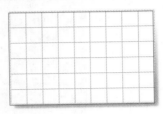

4 acute

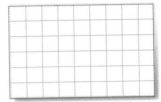

Identify the relationships.

5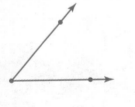

The lines are _____ lines.

6

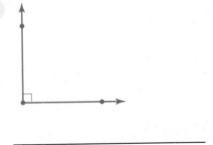

The lines are _____ lines.

Identify each angle.

7

_____

8

_____

Solve.

9 **TIME** Is the measure of the angle formed by the hands of the clock greater than, less than, or equal to 90 degrees?

_____

Two-Dimensional Figures

KEY Concept

Polygons can be named by the number of sides. Some common polygons are shown in the table below.

Polygon	Examples	Sides	Angles
triangle		3	3
quadrilateral		4	4
pentagon		5	5
hexagon		6	6
octagon		8	8

Two-dimensional figures are flat. They do not have thickness. Two-dimensional figures are sometimes called plane figures.

Polygons are two-dimensional figures formed by line segments. Polygons are closed figures. These figures are not polygons.

VOCABULARY

angle
a figure formed when two lines or line segments meet or cross

polygon
a closed plane figure formed by using line segments that meet only at their endpoints

side
one of the line segments that make up a figure

two-dimensional figures
figures that are flat; also called plane figures

GO ON

Example 1

Determine if the figure is a polygon.

1. Is the figure flat? no

2. Is the figure formed by using line segments? yes

3. Is the figure closed? yes

4. If the answers to all the questions are yes, the figure is a polygon.

   This figure is not a polygon.

Example 2

Name the polygon by its number of sides.

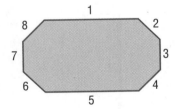

1. Count the number of sides.

2. The polygon has 8 sides.

3. A polygon with 8 sides is called a(n) octagon.

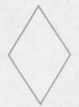
Copyright © by The McGraw-Hill Companies, Inc.

Example 3

Draw a triangle.

1. A triangle has 3 sides formed by line segments.
2. The figure has 3 angles.
3. The figure is closed.

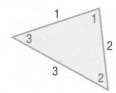

4. The drawing is a triangle.

YOUR TURN!

Draw a pentagon.

1. A _____ has _____ sides formed by line segments.
2. The figure has _____ angles.
3. The figure is _____.

4. The drawing is a _____.

Who is Correct?

Draw a hexagon.

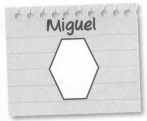

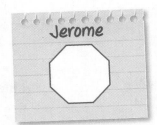

Circle correct answer(s). Cross out incorrect answer(s).

▶ Guided Practice

Determine if each figure is a polygon.

1 _____

2 _____

GO ON

Determine if each figure is a polygon.

3 _____

4 _____

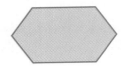

Step by Step Practice

5 **Draw an octagon.**

Step 1 An octagon has _____ sides formed by line segments.

Step 2 The figure has _____ angles.

Step 3 The figure is _____.

Step 4 The drawing is an _____.

Draw each figure.

6 quadrilateral

7 hexagon

Step by Step Problem-Solving Practice

Solve.

8 **SAFETY** Michael saw a street sign that looked like the one at the right. Name the shape of the street sign.

Understand	Read the problem. Write what you know. Michael saw _____.
Plan	Pick a strategy. One strategy is to use a diagram.
Solve	Look at the diagram. Count the number of sides. The street sign has _____ sides. A polygon with _____ sides is called a(n) _____.
Check	Review the definition of a(n) _____.

Problem-Solving Strategies

✓ Use a diagram.

☐ Look for a pattern.

☐ Guess and check.

☐ Solve a simpler problem.

☐ Work backward.

YIELD

9 SIGNS   Anna saw a stop sign like the one at the right. Name the shape of the stop sign.

——— Understand: I circled key words.

——— Plan: To solve the problem, I will _____ .

——— Solve: The answer is _____ .

——— Check: I will check my answer by _____ .

10 TIME   Brandon and his brother play checkers on rainy days. Name the shape of the gameboard.

_____

11 Reflect   Is it possible for a polygon to have only two sides? Explain.

_____

_____

▶ Skills, Concepts, and Problem Solving

Determine if each figure is a polygon.

12 _____

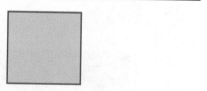

13 _____

14 _____

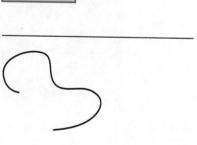

15 _____

Draw each figure.

16 triangle

17 pentagon

GO ON

Solve.

18 **FLAGS**   Hugh looked through a book of state facts. He found the page about Ohio and noticed the flag. What shape is the blue section of the Ohio flag? What shape is the flag? _____

19 **SIGNS**   Akiko and her mother were driving to the airport. Akiko saw a sign like the one at the right. What is the shape of the sign? _____

Vocabulary Check   **Write the vocabulary word that completes each sentence.**

20 *Plane figure* is another term used to name a _____.

21 A(n) _____ is a closed plane figure formed by using line segments that meet only at their endpoints.

22 A polygon with six sides and six angles is called a(n) _____.

Writing in Math   Name some examples of items in your classroom that are shaped like quadrilaterals.

_____

_____

 Spiral Review

Solve. (Lesson 8-1, p. 310)

23 **TRAINS**   Every day Jasmine's school bus driver stops before crossing the train tracks. What kinds of lines are formed by train tracks? _____

Identify each type of angle. (Lesson 8-2, p. 315)

24 _____

25 _____

Length

KEY Concept

Use a **ruler** to measure **length**.

Customary Units of Length

Many people in the United States use customary units of length. Inches, feet, yards, and miles are customary units of length.

The carrot below is measured using a customary ruler. The carrot is $2\frac{1}{2}$ inches in length.

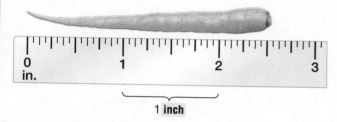

1 **inch**

Metric Units of Length

People around the world use metric units of length. Scientist use the metric system as their standard unit of measurement. Centimeters, millimeters, meters, and kilometers are metric units of length.

The carrot below is measured using a metric ruler.

The carrot is 6.5 centimeters in length.

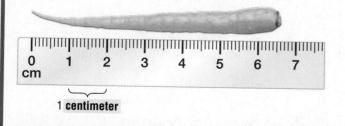

1 **centimeter**

VOCABULARY

centimeter
a metric unit for measuring length

inch
a customary unit for measuring length

length
a measurement of the distance between two points

ruler
a measuring tool used to find the length of an object

GO ON

Example 1

Find the length of the pencil to the nearest inch.

1. Use an inch ruler. Line up the zero mark of the ruler with the left end of the pencil.

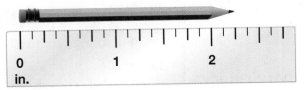

2. Read the number on the ruler that lines up with the right end of the pencil.

The pencil is about 2 inches long.

YOUR TURN!

Find the length of the pen to the nearest inch.

1. Use an inch ruler. Line up the zero mark of the ruler with the left end of the pen.

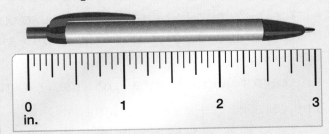

2. Read the number on the ruler that lines up with the right end of the pen.

The pen is about _____ inches long.

Example 2

Find the length of the golf tee to the nearest centimeter.

1. Use a centimeter ruler. Line up the zero mark of the ruler with the left end of the golf tee.

2. Read the number on the ruler that lines up with the right end of the golf tee.

The golf tee is about 5 centimeters long.

YOUR TURN!

Find the length of the paper clip to the nearest centimeter.

1. Use a centimeter ruler. Line up the zero mark of the ruler with the left end of the paper clip.

2. Read the number on the ruler that lines up with the right end of the paper clip.

The paper clip is about _____ centimeters long.

Who is Correct?

Find the length of the yarn to the nearest centimeter.

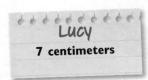  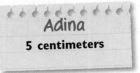

Lucy — 7 centimeters   Brian — 6 centimeters   Adina — 5 centimeters

Circle correct answer(s). Cross out incorrect answer(s).

▶ Guided Practice

Draw a line segment of each length.

1 10 centimeters

2 7 centimeters

3 5 inches

4 2 inches

GO ON

5 Find the height of the mug to the nearest inch.

Step 1 Use an inch ruler. Line up the zero mark of the ruler with the _____ of the mug.

Step 2 Read the number on the ruler that lines up with the _____ of the mug.

The mug is about _____ inches tall.

Measure the length of each line segment to the nearest centimeter.

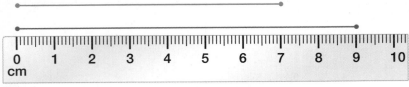

Ruler 1

6 The blue line segment above Ruler 1 is _____ centimeters in length.

7 The red line segment above Ruler 1 is _____ centimeters in length.

Measure the length of each line segment to the nearest inch.

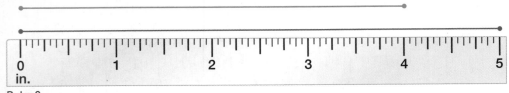

Ruler 2

8 The blue line segment above Ruler 2 is _____ inches in length.

9 The red line segment above Ruler 2 is _____ inches in length.

Step by Step Problem-Solving Practice

Solve.

Problem-Solving Strategies
- ☐ Look for a pattern.
- ☐ Guess and check.
- ☑ Act it out.
- ☐ Solve a simpler problem.
- ☐ Work backward.

10 BASEBALL Charo has a model of a baseball bat. What is the length of the model to the nearest inch?

Understand	Read the problem. Write what you know. Measure the length to the nearest _____.
Plan	Pick a strategy. One strategy is to act it out. Line up the 0 on an inch ruler with the bat.
Solve	Read the closest number on the ruler that lines up with the right end of the baseball bat. The baseball bat is about _____ inches long.
Check	The baseball bat is greater than 4 inches and less than 5 inches long. The answer makes sense.

11 MODELS The post office has a model of a flagpole. What is the height to the nearest centimeter? Check off each step.

_____ Understand: I circled key words.

_____ Plan: To solve the problem, I will _____.

_____ Solve: The answer is _____.

_____ Check: I will check my answer by _____

_____.

12 SHOPPING Ronika bought a ribbon for a dress. What is the length of the ribbon to the nearest inch? _____

GO ON

13 **Reflect** How would you explain to someone how to measure the length of a cell phone?

_____

_____

_____

▶ Skills, Concepts, and Problem Solving

Draw a line segment of each length.

14 4 centimeters

15 6 inches

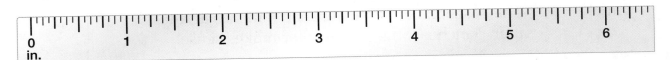

Measure the length of each line segment to the nearest centimeter.

16 The line segment is about _____ centimeters in length.

17 The line segment is about _____ centimeters in length.

Measure the length of each line segment to the nearest inch.

18 The red line segment at the right is about _____ inch(es) tall.

19 The blue line segment at the right is about _____ inch(es) tall.

20 The green line segment at the right is about _____ inch(es) tall.

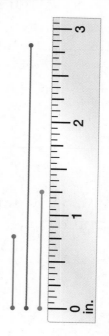

21 The blue line segment below is about _____ inches long.

Solve.

22 TRAVEL   A snail traveled 5 centimeters in an hour. Draw a line segment to show this distance.

23 SCHOOL   After Leo sharpened his pencil, it was 3 inches long. Draw a line segment to show the length of Leo's pencil.

GO ON

Vocabulary Check   **Write the vocabulary word that completes each sentence.**

24 _____ is the distance between two points.

25 A(n) _____ is a measuring tool used to find the length of an object.

26 A(n) _____ is a metric unit for measuring length.

27 **Writing in Math**   Explain how to measure a line segment to the nearest centimeter.

_____

▶ Spiral Review

Solve. (Lesson 8-3, p. 321)

28 **FLAGS**   Rolando and his family are moving to Delaware. He sees a picture of the Delaware flag on a Web site. What shape is the yellow section of the flag?

_____

DECEMBER 7, 1787

29 **PAINTING**   Kayla is painting a design in art class. She makes a shape that has 6 sides. What shape is Kayla's design?

_____

Use grid paper to draw each angle. (Lesson 8-2, p. 315)

30 right angle

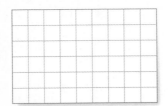

31 straight angle

STOP

Perimeter

KEY Concept

To find the **perimeter** of any **polygon**, add the lengths of all the sides of the polygon.

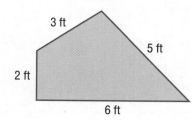

The perimeter of the polygon is $3 + 5 + 6 + 2 = 16$ feet.

VOCABULARY

perimeter
the distance around a shape or region

polygon
a closed plane figure formed using line segments that meet only at their endpoints ⬠

rectangle
a quadrilateral with four right angles; opposite sides are parallel and equal in length ▭

square
a rectangle with four equal sides ◻

triangle
a polygon with three sides and three angles △

Example 1

Find the perimeter of the rectangle.

Add the lengths of the sides.

$7 + 4 + 7 + 4 = 22$

The perimeter is 22 inches.

Check your answer. Use the formula for the perimeter of a rectangle.

$(2 \times \text{length}) + (2 \times \text{width})$
$(2 \times 7) \qquad + (2 \times 4) = 14 + 8 = 22$

YOUR TURN!

Find the perimeter of the rectangle.

Add the lengths of the sides.

$5 + 8 + 5 + 8 =$ _____

The perimeter is _____ inches.

Check your answer. Use the formula for the perimeter of a rectangle.

$(2 \times \text{length}) + (2 \times \text{width})$
$(2 \times 5) \qquad + (2 \times 8) =$ _____ $+$ _____ $=$ _____

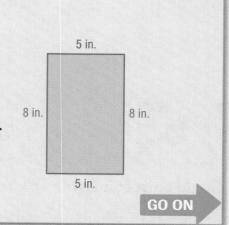

GO ON

Example 2

Find the perimeter of the polygon in centimeters.

1. Measure the length of each side using a centimeter ruler.

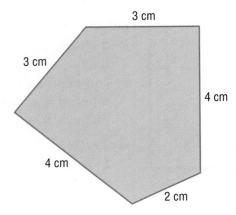

2. Add the lengths of the sides of the polygon.

   $4 + 2 + 4 + 3 + 3 = 16$

   The perimeter of the polygon is 16 centimeters.

YOUR TURN!

Find the perimeter of the polygon in centimeters.

1. Measure the length of each side using a centimeter ruler.

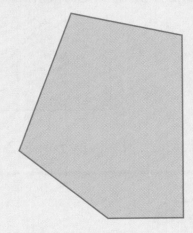

2. Add the lengths of the sides of the polygon.

   _____ + _____ + _____ + _____ + _____ = _____

   The perimeter of the polygon is _____ centimeters.

Who is Correct?

Find the perimeter of the triangle.

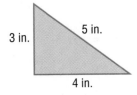

3 in.　5 in.　4 in.

Amy	Keanu	Alma
$5 \times 4 = 20$	$3 + 5 + 4 + 3 = 15$	$3 + 5 + 4 = 12$
20 inches	15 inches	12 inches

Circle correct answer(s). Cross out incorrect answer(s).

Guided Practice

1 Draw a figure that has a perimeter
of 9 centimeters.

2 Draw a figure that has a perimeter
of 10 inches.

Step (by) **Step Practice**

3 Find the perimeter of the polygon in centimeters.

   Step 1 Measure the length of each side
   using a centimeter ruler.

   Step 2 Add the lengths of the sides.

   _____ + _____ + _____ +

   _____ + _____ = _____

   The perimeter of the polygon is _____ centimeters.

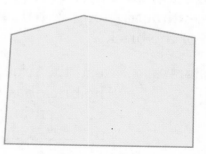

Find the perimeter of each polygon.

4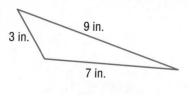

3 in.
9 in.
7 in.

_____ + _____ + _____ = _____

The perimeter of the
triangle is _____ inches.

5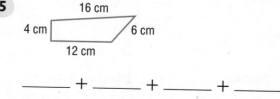

16 cm
4 cm   6 cm
12 cm

_____ + _____ + _____ + _____
= _____

The perimeter of the polygon
is _____ centimeters.

 GO ON

Find the perimeter of each polygon.

6 _____ + _____ + _____ + _____ = _____

The perimeter of the square is _____ inches.
Check your answer.
Perimeter of a square = 4 × length

4 × _____ = _____

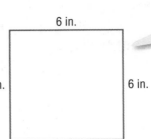
6 in.
6 in.   6 in.
6 in.

> The perimeter of a square can be found by multiplying the length of a side by 4 because all 4 sides of a square are equal.

7 _____ + _____ + _____ + _____ = _____

The perimeter of the rectangle is _____ centimeters.
Check your answer.
Perimeter of a rectangle = (2 × length) + (2 × width)

2 × _____ + 2 × _____ =

_____ + _____ = _____

15 cm
10 cm   10 cm
15 cm

Step by Step Problem-Solving Practice

Problem-Solving Strategies
☑ Draw a diagram.
☐ Look for a pattern.
☐ Guess and check.
☐ Solve a simpler problem.
☐ Work backward.

8 BUILDING Cameron found a brick. Two sides of the top are 19 centimeters long. Two sides are 10 centimeters long. What is the perimeter of the top of the brick?

Understand Read the problem. Write what you know.
The brick has two sides that are
_____ centimeters long and two sides that
are _____ centimeters long.

Plan Pick a strategy. One strategy is to draw a diagram.

Solve _____ the lengths of the sides to find the perimeter.

_____ + _____ + _____ + _____ = _____

Check Use the formula (2 × length) + (2 × width) = perimeter of a rectangle.

(2 × _____) + (2 × _____) =

_____ + _____ = _____

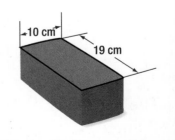
10 cm
19 cm

9 PATIOS   The photo on the right side shows that one side of Gwen's patio measures 11 feet. What is the perimeter of Gwen's patio? Check off each step.

_____ Understand: I circled the key words.

_____ Plan: To solve the problem, I will _____.

_____ Solve: The answer is _____.

_____ Check: I checked my answer by _____.

PATIOS Gwen's patio has the shape of a square.

10 HOBBIES   Each piece of wood that Emil used to build a tree house was 30 inches long and 6 inches wide. What was the perimeter of each piece of wood? _____

11 Reflect   You can find the perimeter of a square by adding the lengths of its sides. Is there another way to find the perimeter of a square? Explain.

_____

_____

> ## Skills, Concepts, and Problem Solving

Draw a polygon that has the given perimeter.

12 20 centimeters

13 19 centimeters

GO ON

Find the perimeter of each polygon.

14 _____ + _____ + _____ = _____

The perimeter of the triangle is
_____ inches.

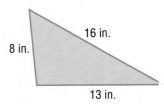

15 _____ + _____ + _____ + _____ = _____

The perimeter of the rectangle is
_____ centimeters.

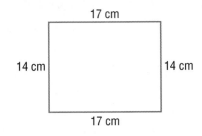

16 _____ + _____ + _____ + _____ + _____ = _____

The perimeter of the polygon is
_____ inches.

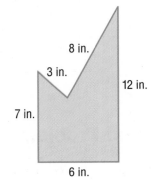

17 _____ + _____ + _____ + _____ = _____

The perimeter of the square is
_____ centimeters.

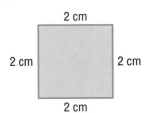

Solve.

18 **ROOMS** Abby's dad is buying baseboard for the living
room. The room is 4 meters long and 7 meters wide. What is
the perimeter of Abby's living room? _____

19 **DESIGN** Rina's mother bought a rug for the entryway. What is the perimeter of the rug? _____

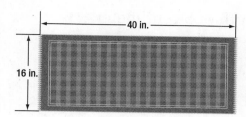

40 in.

16 in.

Vocabulary Check **Write the vocabulary word that completes each sentence.**

20 The _____ is the distance around a shape or region.

21 **Writing in Math** Explain how to find the perimeter of a rectangle that is 9 inches long and 3 inches wide.

_____

_____

▶ Spiral Review

Solve. (Lesson 8-4, p. 327)

22 **TEMPERATURE** What is the length of the thermometer to the nearest inch?

_____

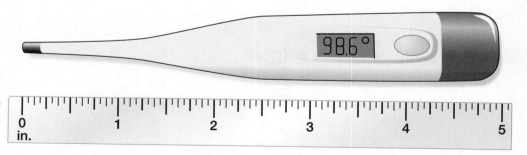

98.6°

0   1   2   3   4   5
in.

Draw each figure. (Lesson 8-1, p. 310)

23 line segment *PQ*

24 line *YZ*

STOP

1 Draw a line segment that has a length of 5 centimeters.

Measure the length of each line segment to the nearest centimeter.

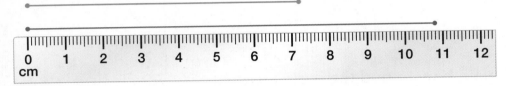

2 Blue line segment: _____

3 Red line segment: _____

Find the perimeter of each polygon.

4 _____

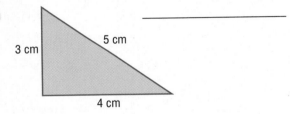

3 cm, 5 cm, 4 cm

5 _____

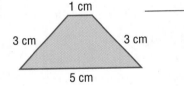

1 cm, 3 cm, 3 cm, 5 cm

Explain if each figure is a polygon.

6 _____
_____

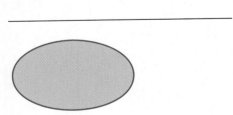

7 _____
_____

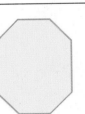

Solve.

8 **SCHOOL** All the desks in Mr. Morgan's class are the same size. Each desk top is 30 inches long and 18 inches wide. What is the perimeter of each desk top? _____

Study Guide

Vocabulary and Concept Check

angle, *p. 315*

centimeter, *p. 327*

degree, *p. 315*

inch, *p. 327*

intersecting lines, *p. 310*

length, *p. 327*

line, *p. 310*

line segment, *p. 310*

parallel lines, *p. 310*

perimeter, *p. 335*

point, *p. 310*

polygon, *p. 321*

rectangle, *p. 335*

ruler, *p. 327*

side, *p. 321*

square, *p. 335*

triangle, *p. 335*

two-dimensional figures, *p. 321*

Write the vocabulary word that completes each sentence.

1 A(n) _____ is a unit for measuring angles.

2 A(n) _____ is a metric unit for measuring length and height.

3 The distance around a shape or a region is its _____.

4 A(n) _____ is a customary unit for measuring length and height.

5 A tool used to measure length is called a(n) _____.

Label the polygon below.

6 _____

7 _____

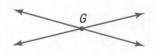

Lesson Review

8-1 Lines (pp. 310–314)

Identify each relationship.

8 _____

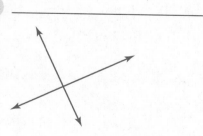

9 _____

Example 1

Identify the relationship.

1. The two lines cross at point G.
2. Lines that cross are called intersecting lines.

8-2 Angles (pp. 315–319)

Use grid paper to draw each angle.

10 acute angle

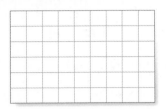

11 right angle

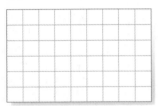

Example 2

Use grid paper to draw an obtuse angle.

1. Draw one line along a line on the grid.

2. The lines of the grid paper intersect at right angles. In order to draw an obtuse angle, draw the second line larger than the sides of one of the squares.

3. Does your angle look larger than a right angle? Yes; it is obtuse.

8-3 Two-Dimensional Figures
(pp. 321–326)

Name each polygon by its number of sides.

12 _____

13 _____

14 _____

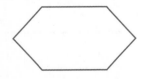

Example 3

Name the polygon by its number of sides.

1. Count the number of sides.

2. The polygon has 5 sides.

3. A polygon with 5 sides is called a(n) pentagon.

8-4 Length (pp. 327–334)

Measure the length of the item to the nearest inch.

15 The key is about _____ long.

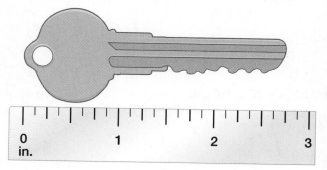

Measure the length of the item.

16 The earthworm is

_____ long.

Example 4

Find the length of the crayon to the nearest inch.

Use an inch ruler. Line up the zero mark of the ruler with the left end of the crayon.

Read the number on the ruler that lines up with the right end of the crayon. The crayon is about 3 inches long.

8-5 Perimeter (pp. 335–341)

Find the perimeter of each polygon.

17

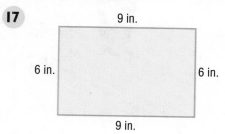

18

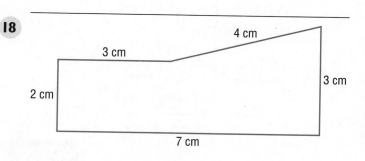

Example 5

Find the perimeter of the polygon in inches.

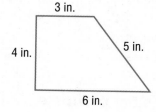

Add the lengths of the sides.
$$4 + 3 + 6 + 5 = 18$$
The perimeter of the polygon is 18 inches.

Draw each figure.

1 intersecting lines

2 line segment *ST*

Identify the relationship.

3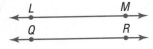

Identify the angle.

4

_____

_____

5 What is the height of the action figure at the right to the nearest centimeter?

_____

6 Draw a line segment that has a length of 5 centimeters.

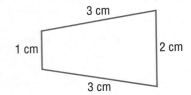

7 What is the perimeter of the polygon? _____

3 cm

1 cm

2 cm

3 cm

8 What is the perimeter of the rectangle? _____

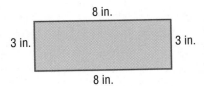

8 in.

3 in.   3 in.

8 in.

GO ON

Determine if the figure is a polygon.

9 _____

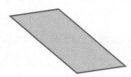

10 _____

Solve.

11 **CARDS** Mrs. Chung has a membership card to the local video store. The card has a width of 5 centimeters. It has a length of 8 centimeters. What is the perimeter of the card? _____

Circle Town Video Rental

Mrs. L. R. Chung

12 **POLYGONS** The membership card is in the shape of a rectangle. How would you describe the line segments that form a rectangle?

_____

_____

13 **CALENDARS** Ms. Morris keeps a calendar on her desk. The photo on the calendar has a perimeter of 36 centimeters. The photo is square. What is the length of each side of the photo on the calendar? Explain your answer.

_____

_____

_____

Correct the mistakes.

14 Kenyi was asked to describe the polygon at the right. She said that it is a quadrilateral with 3 obtuse angles. Explain Kenyi's mistake(s).

_____

_____

_____

STOP

Choose the best answer and fill in the corresponding circle on the sheet at right.

1 Find the length of the line segment to the nearest centimeter.

A 2 cm   C 7 cm

B 5 cm   D 10 cm

2 What is the perimeter of the rectangle?

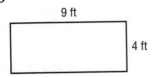

9 ft

4 ft

A 13 feet   C 36 feet

B 26 feet   D 42 feet

3 What is the perimeter of the triangle?

14 in.   14 in.

14 in.

A 14 inches   C 42 inches

B 28 inches   D 196 inches

4 Larisa has 3 dominoes. Each domino is 4 centimeters long. How many centimeters long is the line of dominoes?

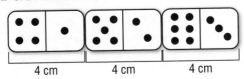

4 cm   4 cm   4 cm

A 7 centimeters   C 12 centimeters

B 10 centimeters   D 24 centimeters

5 Which best describes these lines?

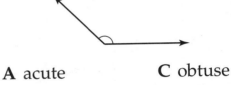

A perpendicular   C bisecting

B intersecting   D parallel

6 What type of angle is shown?

A acute   C obtuse

B right   D straight

7 Find the length of the line segment to the nearest centimeter.

A 3 cm

B 5 cm

C 8 cm

D 12 cm

8 Walter wants to show his friends an example of a right angle. What should he draw?

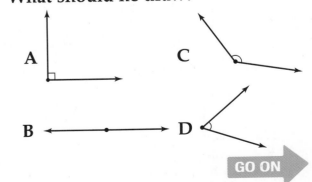

A

B

C

D

GO ON

9 What is the name of the figure?

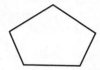

    A line segment *ST* **C** point *ST*

    B line *ST*   **D** quadrilateral

---

10 What is the name of the figure?

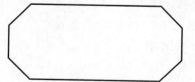

    A quadrilateral   **C** hexagon

    B pentagon   **D** octagon

---

11 Find the length of the line segment to the nearest inch.

    A 3 inches   **C** 6 inches

    B 5 inches   **D** 7 inches

---

12 What is the name of the figure?

    A quadrilateral   **C** hexagon

    B pentagon   **D** octagon

ANSWER SHEET

Directions: Fill in the circle of each correct answer.

1   (A) (B) (C) (D)

2   (A) (B) (C) (D)

3   (A) (B) (C) (D)

4   (A) (B) (C) (D)

5   (A) (B) (C) (D)

6   (A) (B) (C) (D)

7   (A) (B) (C) (D)

8   (A) (B) (C) (D)

9   (A) (B) (C) (D)

10   (A) (B) (C) (D)

11   (A) (B) (C) (D)

12   (A) (B) (C) (D)

Success Strategy

If you do not know the answer to a question, go on to the next question. Come back to the problem, if you have time. You might find another question later in the test that will help you figure out the skipped problem.

STOP

Area

How do you use area?

You can use area to decorate your room. Knowing the area of the floor will help you choose the best rug. Finding the area of a wall will help you buy enough paint.

STEP 1 Quiz

Are you ready for Chapter 9? Take the Online Readiness Quiz at *glencoe.com* to find out.

STEP 2 Preview

Get ready for Chapter 9. Review these skills and compare them with what you will learn in this chapter.

What You Know	What You Will Learn
You can identify geometric figures. square	*Lesson 9-1* You will learn how some figures can be used to make other figures. 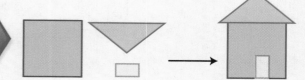
TRY THIS! **Identify each figure.** 1 _____ 2 ▽ _____	
You know that two halves equal one whole. $\frac{1}{2} + \frac{1}{2} = 1$	*Lesson 9-2* You can count square units to find area. 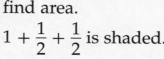 $1 + \underbrace{\frac{1}{2} + \frac{1}{2}}$ is shaded. $1 + \quad 1 \quad = 2$ square units
You know how to measure the lengths of items. 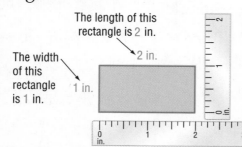	*Lesson 9-3* To find the area of a rectangle, multiply the length by the width. Area $=$ length $\times$ width Area $= 2$ inches $\times 1$ inch Area $= 2$ square inches

Create Figures

KEY Concept

You can put figures together to create new figures.

3 rectangles become 1 rectangle.

You can cut figures into parts to create new figures.

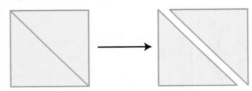

1 square becomes 2 triangles.

VOCABULARY

hexagon
a polygon with 6 sides and 6 angles

pentagon
a polygon with 5 sides

Example 1

Use triangles to make a hexagon.

1. A hexagon has 6 sides.

2. Arrange the triangles into a hexagon. Make sure that the figures do not overlap.

3. Trace or draw the hexagon you made.

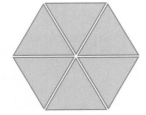

YOUR TURN!

Use triangles to make a larger triangle.

1. A triangle has _____ sides.

2. Arrange the triangles into a larger triangle.

3. Trace or draw the triangle you made.

Example 2

Draw lines in the square to make 8 rectangles.

1. A square has 4 sides.

2. Draw lines in the square to create 4 squares.

3. Keep drawing lines until there are 8 rectangles.

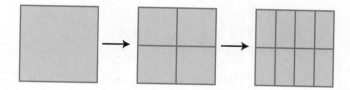

YOUR TURN!

Draw 3 lines in the triangle to make 6 triangles.

1. A triangle has _____ sides.

2. Draw 1 line in the triangle to create 2 triangles.

3. Keep drawing lines until there are 6 triangles.
   (Hint: Try drawing a line from each of the vertices (corners).

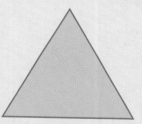

GO ON

Who is Correct?

Suppose the two triangles shown were put together. Would they make a square, a rectangle, or a hexagon?

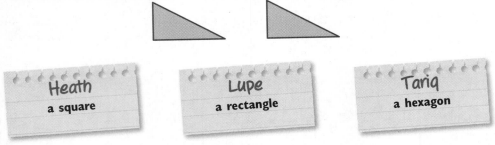

Circle correct answer(s). Cross out incorrect answer(s).

▶ Guided Practice

Find the figures. Write the letters.

1 Find 2 squares. _____

2 Find 2 triangles. _____

3 Find 1 hexagon. _____

4 Find 1 rectangle. _____

Create each figure.

5 Use 3 triangles to make a rectangle. Trace or draw the rectangle you made.

6 Use at least 2 different shapes to create a rectangle. Trace or draw the shape you made.

Step by Step Practice

Create shapes.

7 Draw 1 line in the triangle to make 2 figures.

Step 1 Draw 1 line in the triangle.

Step 2 Identify the 2 figures.

_____

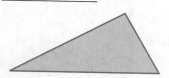

Create figures.

8 Draw 2 lines in the square to make 4 squares.

9 Draw 5 lines in the rectangle to make 6 triangles.

Decide if each situation describes building figures or breaking them apart. Write *break apart* or *build*.

10 ART Katie wants small triangles of paper for an art project. She has a large hexagon piece of paper. Katie must _____ the hexagon.

11 QUILTS Dalila is making a quilt design. She wants a large red square in the quilt. Dalila has red triangle pieces. She must _____ the large square with the red triangles.

Step by Step Problem-Solving Practice

Problem-Solving Strategies

- ☑ Use a diagram.
- ☐ Look for a pattern.
- ☐ Guess and check.
- ☐ Act it out.
- ☐ Solve a simpler problem.

12 **ARCHITECTURE** Identify three different figures in the building.

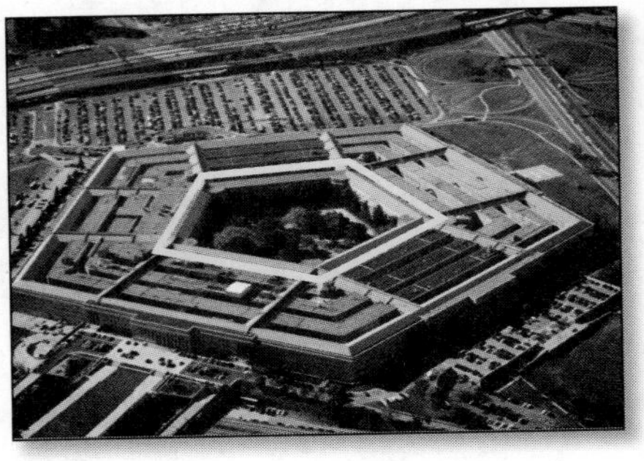

Understand Read the problem. Write what you know. You need to identify ________________ in the building.

Plan Pick a strategy. One strategy is to use a diagram. Draw lines on the picture to identify and count the figures.

Solve List the figures.

Figure 1: ________________

Figure 2: ________________

Figure 3: ________________

Check Count the figures that you drew to make sure that you found three different figures.

13 **COOKING** Santos made a square pizza. It would not fit in his oven, so he cut it into two equal pieces. What two figures could the pieces be?

Check off each step.

_____ **Understand: I circled key words.**

_____ **Plan: To solve the problem, I will** ________________.

_____ **Solve: The answer is** ________________.

_____ **Check: I will check my answer by** ________________.

14 **HOBBIES** Yoshime made a kite frame out of straws. She needs to cover the frame with paper. She has three square-shaped pieces of paper that will cover the kite. What will be the final figure?

____

15 **Reflect** Think about what you know about creating figures. Why is it important that the figures do not overlap?

_____

_____

▶ Skills, Concepts, and Problem Solving

16 Use 4 figures to create a hexagon. Trace or draw the hexagon you made.

17 Use 4 squares to make a rectangle. Trace or draw the rectangle you made.

18 Draw 1 line in the triangle to make 2 new figures.

19 Draw 2 lines in the rectangle to make 3 new figures.

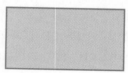

20 Draw 3 lines in the rectangle to make at least 4 new figures.

21 Draw 4 lines in the square to make 9 smaller squares.

GO ON

Solve.

22 **RECREATION**   Identify three different figures in the playground structure.

_____

23 **INTERIOR DESIGN**   Kenyon is painting his bedroom. One wall is a square. He wants to make four smaller squares on the wall. How can he make four squares?

_____

_____

24 **GEOMETRY**   Diana is using paper shapes to do a geometry project. The directions say to paste two triangles on a piece of paper. Diana only has one rectangle. How can she make two triangles?

_____

25 **PUZZLES**   Logan has a picture of the ocean. He wants to cut it into a simple puzzle for his younger sister. Draw lines on the picture to show how Logan could cut the picture into eight pieces.

Vocabulary Check   **Write the vocabulary word or words that complete each sentence.**

26 A(n) _____ is a polygon with five sides.

27 A polygon with 6 sides is called a(n) _____.

28 **Writing in Math**   Describe how to use lines to cut a rectangle into 3 or more smaller rectangles.

_____

_____

_____

STOP

Introduction to Area

KEY Concept

The **area** of a figure is the number of **square units** needed to cover a surface.

To find the area of a figure, you can count the number of square units the figure covers.

1	2	3	4
5	6	7	8
9	10	11	12
13	14	15	16
17	18	19	20

The area of the rectangle is 20 square units.

VOCABULARY

area
the number of square units needed to cover a region or plane figure

square unit
a unit for measuring area

The units of area are square units.

Example 1

Find the area of the rectangle.

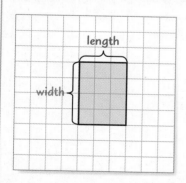

Count the number of squares the rectangle covers.

The area is 12 square units.

YOUR TURN!

Find the area of the rectangle.

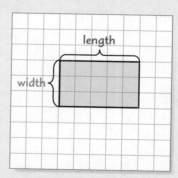

Count the number of squares the rectangle covers.

The area is _____ square units.

GO ON

Example 2

Estimate the area of the figure.

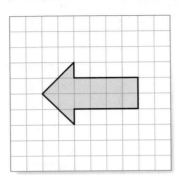

1. Count the number of whole squares the figure covers.
   10 whole squares

2. Count the number of half squares the figure covers.
   4 half squares

   Convert the half squares to whole squares.
   $\frac{1}{2} + \frac{1}{2} + \frac{1}{2} + \frac{1}{2} = 2$

3. Add the number of whole squares.
   10 + 2 = 12

The area of the figure is about 12 square units.

YOUR TURN!

Estimate the area of the figure.

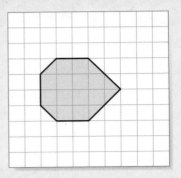

1. Count the number of whole squares the figure covers.

   _____ whole squares

2. Count the number of half squares the figure covers.

   _____ half squares = _____ whole square(s)

3. Add the number of whole squares.

   _____ + _____ = _____

The area of the figure is about

_____ square units.

Who is Correct?

Find the area of the square.

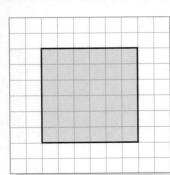

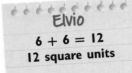

Darcy
A square has 4 sides.
4 square units

Elvio
6 + 6 = 12
12 square units

Ian
6 × 6 = 36
36 square units

Circle correct answer(s). Cross out incorrect answer(s).

▶ Guided Practice

Draw a figure that has the given area.

1 28 square units

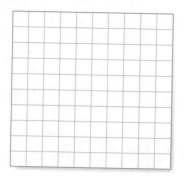

2 16 square units

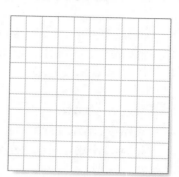

Step (by) Step Practice

3 Estimate the area of the figure.

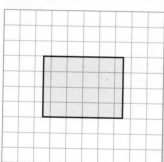

Step 1 Count the number of whole squares the figure covers.

_____ whole squares

Step 2 Count the number of half squares.

_____ half squares = _____ whole square(s)

Step 3 Add the number of whole squares.

_____ + _____ = _____

The area of the figure is about

_____ square units.

Find the area of each figure.

4

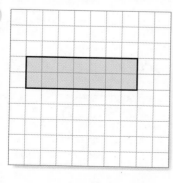

_____ square units

5

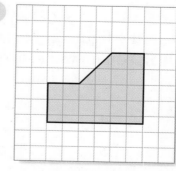

_____ square units

GO ON ➡

Step by Step Problem-Solving Practice

Problem-Solving Strategies

- ☑ Draw a diagram.
- ☐ Look for a pattern.
- ☐ Guess and check.
- ☐ Act it out.
- ☐ Solve a simpler problem.

Solve.

6 **ART** Gregory's painting measures 8 inches by 10 inches. What is the area of the painting?

Understand Read the problem. Write what you know. A painting has sides of ______ units and ______ units.

Plan Pick a strategy. One strategy is to draw a diagram. Draw a rectangle that has sides of 8 units and 10 units.

Solve Count the number of squares the figure covers. The area of the rectangle is ______ square units.

Check Add the number of squares in each row.

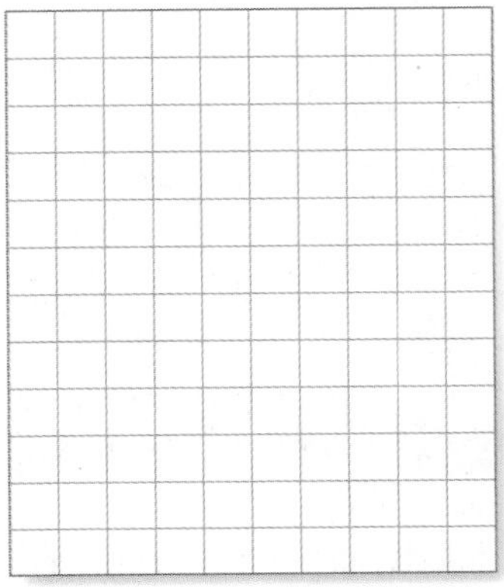

7 **ROOMS** Selina's bedroom floor measures 12 feet by 8 feet. What is the area of the floor? Check off each step.

______ **Understand: I circled key words.**

______ **Plan: To solve the problem, I will** ______________________.

______ **Solve: The answer is** ______________________.

______ **Check: I checked my answer by** ______________________.

8 **INTERIOR DESIGN** James bought a poster that was 2 feet wide and 4 feet long. What was the area of the poster?

____

9 **Reflect** Look at the figure at the right. Is the area about 15 square units? Explain.

____

____

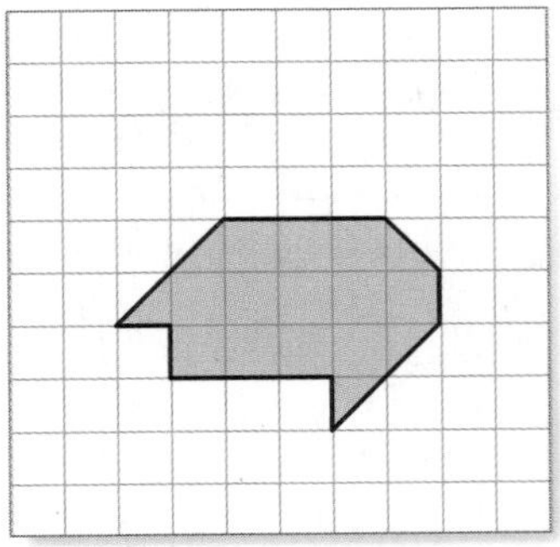

Skills, Concepts, and Problem Solving

Draw a figure that has the given area.

10 23 square units

11 15 square units

Find the area of each figure.

12

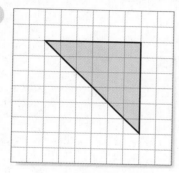

_____ square units

13

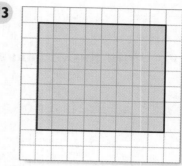

_____ square units

14

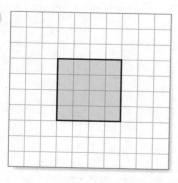

_____ square units

15

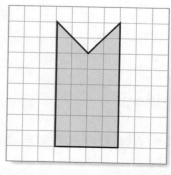

_____ square units

GO ON

Solve.

16 GARDENS   Ling planted a rectangular garden. Two sides were 6 feet long. Two sides were 8 feet long. What was the area of Ling's garden?

_____

17 GAMES   Gabriel's favorite music video game uses a dance pad. The pad has 6 squares across and 3 squares down. What is the area of the dance pad?

_____

18 INTERIOR DESIGN   Tanya's rug is 3 feet wide. It is 5 feet long. What is the area of the rug?

_____

Vocabulary Check   **Write the vocabulary word that completes each sentence.**

19 A(n) _____ is a unit for measuring area.

20 _____ is the number of square units needed to cover a region or plane figure.

21 Writing in Math   Explain how to find the length and width of a rectangle with an area of 48 square units.

_____

_____

▶ **Spiral Review**

Find the figures. Write the letters. (Lesson 9-1, p. 352)

22 Find 1 rectangle. _____

23 Find 4 triangles. _____

24 Find 2 squares. _____

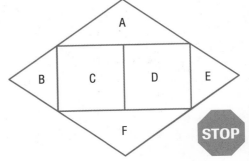

Create each figure.

1 Use 2 triangles to create a rectangle. Trace or draw the rectangle you made.

2 Draw 2 lines on the rectangle to make 3 triangles.

Find the area of each figure.

3 The area of the rectangle is

_____.

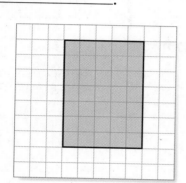

4 The area of the figure is about

_____.

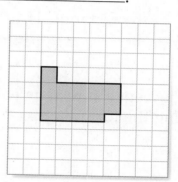

Solve.

5 **COOKING** Kelly made a rectangular pan of brownies. She wants to cut them into squares. She wants 12 squares in all. How many lines could she cut in the brownies to make 12 squares?

_____

6 **INTERIOR DESIGN** The ceiling tiles in Albert's kitchen are square. The room is 12 feet by 16 feet. What is the area of the ceiling?

_____

Area of a Rectangle

KEY Concept

Find the **area** of a **rectangle** using the formula below.

ℓ is the length of
the rectangle.

A is the area of
the rectangle. $\longrightarrow A = \ell \times w \longleftarrow$ w is the width of the
rectangle.

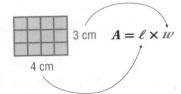 3 cm $A = \ell \times w$

4 cm

The area of the rectangle is 12 square centimeters.

A **square** is a special rectangle. Find the area of a
square using the formula $A = \ell \times w$.

VOCABULARY

area
the number of square
units needed to cover
a region or a plane
figure

rectangle
a quadrilateral with
four right angles;
opposite sides are
parallel and equal in
length

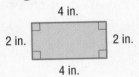

4 in.

2 in.   2 in.

4 in.

square
a rectangle with equal
sides

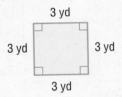

3 yd

3 yd   3 yd

3 yd

square unit
a unit for measuring
area

Example 1

Find the area of the rectangle.

1. The length is 5 inches.
   The width is 3 inches.

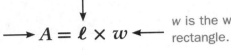

width =
3 in.

length = 5 in.

2. Substitute these values into
   the formula. Multiply.

   $A = \ell \times w$
   $A = 5 \text{ in.} \times 3 \text{ in.}$
   $A = 15 \text{ in}^2$

The area of the rectangle is 15 square inches.

YOUR TURN!

Find the area of the rectangle.

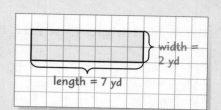

1. The length is _____ yards.
   The width is _____ yards.

2. Substitute these values into the formula. Multiply.

   $A = \ell \times w$

   $A =$ _____ yd $\times$ _____ yd

   $A =$ _____ yd²

The area of the rectangle is _____ square yards.

Example 2

Find the area of the square.

1. The length is 4 feet.
   The width is 4 feet.

2. Substitute these values into the formula. Multiply.

   $A = \ell \times w$
   $A = 4 \text{ ft} \times 4 \text{ ft}$
   $A = 16 \text{ ft}^2$

The area of the square is 16 square feet.

YOUR TURN!

Find the area of the square.

1. The length is
   _____ kilometers.
   The width is
   _____ kilometers.

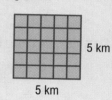

2. Substitute these values into the formula. Multiply.

   $A = \ell \times w$

   $A =$ _____ km $\times$ _____ km

   $A =$ _____ km²

The area of the square is
_____ square kilometers.

Who is Correct?

What is the area of the rectangle at the right?

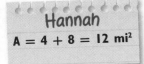

Hannah
A = 4 + 8 = 12 mi²

Ashanti
A = 4 × 8 = 32 mi²

Benito
A = 8 × 8 = 64 mi²

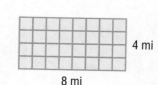

Circle correct answer(s). Cross out incorrect answer(s).

GO ON

▶ Guided Practice

Draw a rectangle that has the given area.

1   4 cm²

2   21 cm²

Step by Step Practice

3   Find the area of the rectangle.

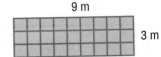

9 m

3 m

   Step 1   The length is _____ meters.
                 The width is _____ meters.

   Step 2   Substitute these values into the formula. Multiply.

$$A = \ell \times w$$
$$A = \underline{\quad} \text{ m} \times \underline{\quad} \text{ m}$$
$$A = \underline{\quad} \text{ m}^2$$

            The area of the rectangle is _____ square meters.

Find the area of each rectangle.

4   The length is _____ feet. The width is _____ feet.

$$A = \underline{\quad} \text{ ft} \times \underline{\quad} \text{ ft}$$
$$A = \underline{\quad} \text{ ft}^2$$

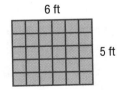

6 ft

5 ft

5   The length is _____ inches. The width is
     _____ inches.

$$A = \ell \times w$$
$$A = \underline{\quad} \text{ in.} \times \underline{\quad} \text{ in.}$$
$$A = \underline{\quad} \text{ in}^2$$

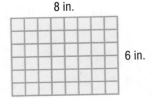

8 in.

6 in.

6 $A =$ _____

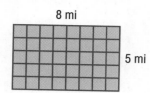

8 mi

5 mi

7 $A =$ _____

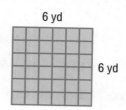

6 yd

6 yd

Step by Step Problem-Solving Practice

Solve.

8 SOCCER A soccer field is 80 yards long and 40 yards wide. What is the area of the field?

Understand Read the problem. Write what you know.

The length is _____ yards. The width is _____ yards.

Plan Pick a strategy. One strategy is to use a formula.

Substitute these values into the area formula.

Solve Use the formula.

$A = \ell \times w$

$A =$ _____ yd $\times$ _____ yd

$A =$ _____ yd²

The area of the soccer field is _____ square yards.

Check Use division or multiplication to check your answer.

GO ON

9 **MOVIES**   The screen in a movie theater is 18 feet long.
It is 12 feet high. What is the area of the movie screen?
Check off each step.

_____ Understand: I circled key words.

_____ Plan: To solve the problem, I will _____.

_____ Solve: The answer is _____.

_____ Check: I checked my answer by _____.

10 **ART**   Mr. Parker asked his students to use a sheet of
paper for a painting. Each sheet of paper measured
11 inches by 14 inches. What was the area of each painting?

_____

11 **Reflect**   Can two rectangles have the same area but
different lengths and widths? Explain.

_____

_____

_____

▶ **Skills, Concepts, and Problem Solving**

Draw a rectangle that has the given area.

12 24 square units

13 6 square units

Find the area of each rectangle.

14 $A =$ _____

12 km

3 km

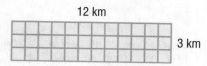

15 $A =$ _____

8 m

2 m

16 $A =$ _____

9 in.

4 in.

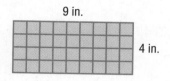

17 $A =$ _____

6 yd

3 yd

18 $A =$ _____

10 cm

10 cm

19 $A =$ _____

7 mi

4 mi

Solve.

20 **INTERIOR DESIGN**   A carpet is 4 meters wide. It is 6 meters long. Find the area of the carpet.

_____

21 **TOYS**   Kara's younger brother has a carrying case for his toy cars. The case fits 6 cars across and 8 cars down. How many cars will fit in all?

_____

22 **GARDENING**   A flower bed is 7 meters long and 3 meters wide. What is the area of the flower bed?

_____

GO ON

23 PHOTOGRAPHY Bianca takes photographs of horses. This photograph is 28 centimeters tall and 36 centimeters wide. What is the area of the photograph?

36 cm
28 cm

_____

Vocabulary Check **Write the vocabulary word that completes each sentence.**

24 _____ is the number of square units needed to cover a region or a plane figure.

25 A(n) _____ is a rectangle with four equal sides.

26 A(n) _____ has opposite sides that are equal and parallel. It is a quadrilateral with four right angles.

27 Writing in Math Explain how to find the area of a square.

_____

_____

_____

 Spiral Review

Find the area of each figure. (Lesson 9-2, p. 359)

28 The area of the square is

_____.

7 ft

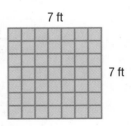

7 ft

29 The area of the rectangle is

_____.

5 in.

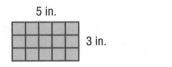

3 in.

Solve.

30 MEASUREMENT Malik wants to measure his bedroom door so he can put up posters. The door is 3 feet wide and 7 feet tall. What is the area of his door?

_____

STOP

Study Guide

Vocabulary and Concept Check

area, *p. 359*
hexagon, *p. 352*
pentagon, *p. 352*
rectangle, *p. 366*
square, *p. 366*
square unit, *p. 359*

Write the vocabulary word that completes each sentence.

1 A(n) _____ is a rectangle with four equal sides.

2 The number of square units needed to cover a region or plane figure is the _____.

3 A(n) _____ is a parallelogram with four right angles.

4 The units used for measuring area is called a(n) _____.

Label each diagram below. Write the correct vocabulary term in each blank.

5 _____

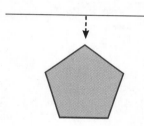

6 _____

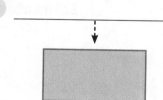

7 _____

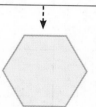

Lesson Review

9-1 Create Figures (pp. 352–358)

Create the figure.

8 Use 2 triangles to make a square. Trace or draw the square you made.

Example 1

Create the figure.

Use 3 squares to make a rectangle.

1. A rectangle has 4 sides.

2. Arrange the squares into a rectangle.

3. Trace or draw the rectangle you made.

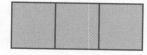

Create the figure.

9 Draw 2 lines in the triangle to make 3 triangles.

Example 2

Draw 2 lines in the rectangle to make 3 figures.

1. A rectangle has 4 sides.

2. Draw 1 line in the rectangle to create 2 figures.

3. Draw another line to create 3 figures.

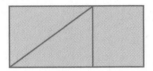

9-2 Introduction to Area (pp. 359–364)

Find the area of each figure.

10

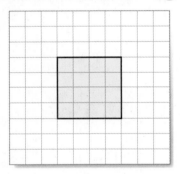

11

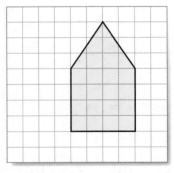

Example 3

Estimate the area of the figure.

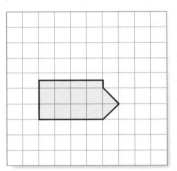

Count the number of whole squares the figure covers. **8 whole squares**

Count the number of half squares the figure covers. 6 half squares

$$\frac{1}{2} + \frac{1}{2} + \frac{1}{2} + \frac{1}{2} + \frac{1}{2} + \frac{1}{2} = 3$$

Add the number of whole squares.
8 + 3 = 11

The area of the figure is 11 square units.

9-3 Area of a Rectangle (pp. 366–372)

Find the area of each rectangle.

12

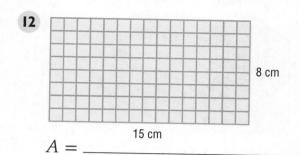

8 cm

15 cm

A = _____

13

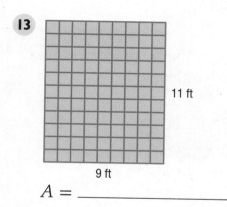

11 ft

9 ft

A = _____

Find the area of each square.

14

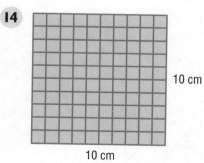

10 cm

10 cm

A = _____

15

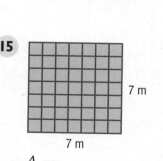

7 m

7 m

A = _____

Example 4

Find the area of the rectangle.

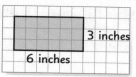

3 inches

6 inches

ℓ is the length of the rectangle.

w is the width of the rectangle.

A is the area of the rectangle.

$$A = \ell \times w$$

The length is 6 inches.
The width is 3 inches.

Substitute these values into the formula. Multiply.

$A = \ell \times w$
$A = 6 \text{ in.} \times 3 \text{ in.}$
$A = 18 \text{ in}^2$

The area of the rectangle is 18 square inches.

Example 5

Find the area of the square.

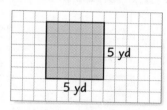

5 yd

5 yd

The length is 5 yards.
The width is 5 yards.

Substitute these values into the formula. Multiply.

$A = \ell \times w$
$A = 5 \text{ yd} \times 5 \text{ yd}$
$A = 25 \text{ yd}^2$

The area of the square is 25 square yards.

Find the figures. Write the letters.

1 Find 3 rectangles. _____

2 Find 1 triangle. _____

3 Find 1 square. _____

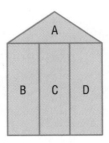

Create figures.

4 Use 4 squares to create a larger square. Trace or draw the square you made.

5 Draw lines in the hexagon to make at least 4 triangles.

Find the area of each figure.

6 _____

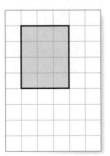

7 _____

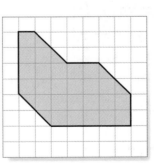

GO ON

Draw a rectangle that has the given area.

8 35 units²

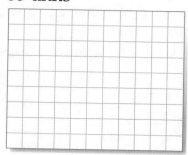

9 49 units²

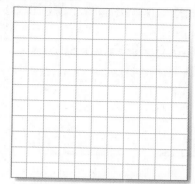

Find the area of each rectangle.

10

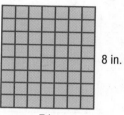

8 in.

7 in.

$A = $ _____

11

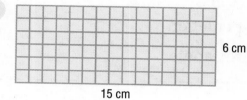

6 cm

15 cm

$A = $ _____

Solve.

12 COOKING Bena made square cookies. One side of each cookie was 5 centimeters. What was the area of each cookie?

_____

13 PETS Max's dog sleeps on a rectangular rug. It is 2 feet wide and 4 feet long. What is the area of the rug?

_____

Correct the mistakes.

14 GAMES Anton drew a 4-square game on the sidewalk. The large square was 4 feet on each side. Anton said that the area was 8 square feet because 4 + 4 = 8. What mistake did he make?

_____

15 Explain how Anton should have found the area.

_____

STOP

Choose the best answer and fill in the corresponding circle on the sheet at right.

1 Suppose the figures were put together. Which shape could they create?

 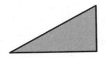

A square

B triangle

C pentagon

D rectangle

---

2 Zoe's poster is 2 feet wide and 3 feet long. What is the area of the poster?

A 5 feet

B 6 ft²

C 8 ft²

D 23 ft²

---

3 How many sides does a hexagon have?

A 3

B 4

C 5

D 6

4 Find the area of the shaded figure.

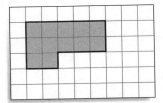

A 12 square units

B 16 square units

C 21 square units

D 28 square units

---

5 Find the area of the rectangle.

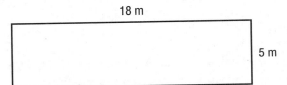

18 m

5 m

A 23 m²

B 46 m²

C 90 m²

D 100 m²

---

6 One wall in Mr. Garcia's classroom measures 18 feet by 10 feet. What is the area of this wall?

A 28 ft²

B 80 ft²

C 180 ft²

D 1,800 ft²

GO ON

7 Find the area of each figure. Which sentence is true?

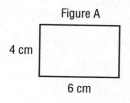
Figure A
4 cm
6 cm

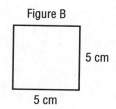
Figure B
5 cm
5 cm

A Area A > Area B

B Area B < Area A

C Area A < Area B

D Area A = Area B

8 Find the area of the square.

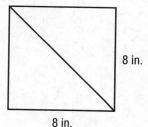
8 in.
8 in.

A 16 in²   **C** 32 in²

B 24 in²   **D** 64 in²

9 Find the area of the rectangle.

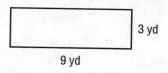
3 yd
9 yd

A 12 yards

B 27 yd²

C 30 yards

D 36 yd²

10 Find the area of the shaded figure.

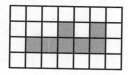

A 7 square units

B 8 square units

C 21 square units

D 28 square units

ANSWER SHEET

Directions: Fill in the circle of each correct answer.

1 Ⓐ Ⓑ Ⓒ Ⓓ
2 Ⓐ Ⓑ Ⓒ Ⓓ
3 Ⓐ Ⓑ Ⓒ Ⓓ
4 Ⓐ Ⓑ Ⓒ Ⓓ
5 Ⓐ Ⓑ Ⓒ Ⓓ
6 Ⓐ Ⓑ Ⓒ Ⓓ
7 Ⓐ Ⓑ Ⓒ Ⓓ
8 Ⓐ Ⓑ Ⓒ Ⓓ
9 Ⓐ Ⓑ Ⓒ Ⓓ
10 Ⓐ Ⓑ Ⓒ Ⓓ

Success Strategy

Try to answer every question. Work out the problem and eliminate answers you know are wrong. Do not change your answers unless you are very uncertain about your first answer choice.

STOP

Spatial Reasoning

A reflection does not change shape or size.

A reflection shows you exactly how you look. You can see your reflection in bodies of water such as pools, ponds, lakes, and oceans. You can also tell exactly how you look by using a mirror.

STEP 1 Quiz

Are you ready for Chapter 10? Take the Online Readiness Quiz at *glencoe.com* to find out.

STEP 2 Preview

Get ready for Chapter 10. Review these skills and compare them with what you will learn in this chapter.

What You Know	What You Will Learn

What You Know

You know how to find matching figures.

The two red circles are a match.

TRY IT!

Draw a circle around the two matching figures.

You know that a butterfly is the same on both sides.

What You Will Learn

Lesson 10-1

Congruent figures have the same size and same shape.

Even if one of the figures is turned in a different direction, they are still congruent.

Lesson 10-3

The line that divides a figure in half so that both parts match is called the **line of symmetry**.

Some figures do not have line symmetry. Some figures have more than one line of symmetry.

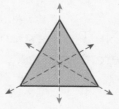

Congruent Figures

KEY Concept

Congruent figures have the same shape and the same size.

Polygons are congruent if their corresponding sides are of equal length. When figures are congruent, they fill the same number of squares on a grid.

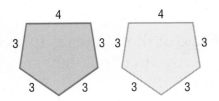

Circles are congruent if they have the same diameter.

Even if figures are turned in different directions, they can still be congruent.

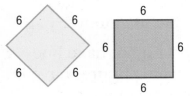 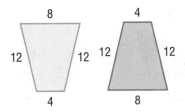

Figures are NOT congruent when they are different sizes or different shapes.

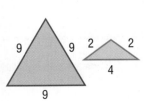 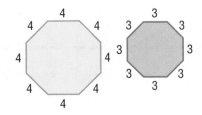

VOCABULARY

congruent figures
figures that are the same size and the same shape

polygon
a closed plane figure formed by using line segments that meet only at their end points

Example 1

Decide if the figures are congruent.

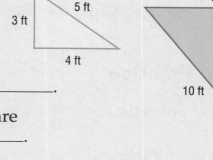

1. The side lengths of the blue quadrilateral are 3 inches, 2 inches, 3 inches, and 2 inches.

2. The side lengths of the yellow quadrilateral are 3 inches, 2 inches, 3 inches, and 2 inches.

3. The quadrilaterals have sides that are the same length.

4. The figures are congruent.

YOUR TURN!

Decide if the figures are congruent.

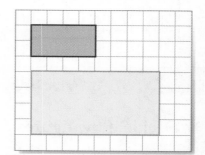

1. The side lengths of the green triangle are _____, _____, and _____.

2. The side lengths of the yellow triangle are _____, _____, and _____.

3. The triangles have sides that are _____.

4. The figures are _____.

Example 2

Decide if the figures are congruent.

1. Compare the two figures. Do they fill the same number of squares on the grid? **no**

2. Are they the same shape? **yes**

3. Are they congruent? **no**

YOUR TURN!

Decide if the figures are congruent.

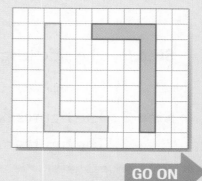

1. Compare the two figures. Do they fill the same number of squares on the grid? _____

2. Are they the same shape? _____

3. Are they congruent? _____

GO ON

Who is Correct?

Decide if the figures are congruent.

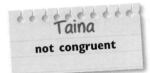

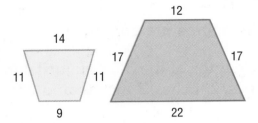

Mallory
congruent

Taina
not congruent

Kevin
similar

Circle correct answer(s). Cross out incorrect answer(s).

▶ Guided Practice

1 Draw two figures that are congruent.

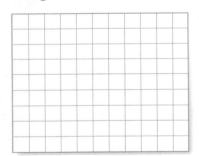

2 Draw two figures that are not congruent.

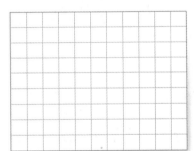

Step (by) Step Practice

3 Decide if the figures are congruent.

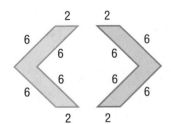

Step 1 The side lengths of the orange figure are:

_____.

Step 2 The side lengths of the green figure are:

_____.

Step 3 The figures have sides that are the same lengths.

The figures are _____.

Decide if the figures are congruent.

4

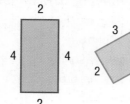

The side lengths of the blue figure are:

_____.

The side lengths of the red figure are:

_____.

The figures have sides that are

_____.

The figures are _____.

5

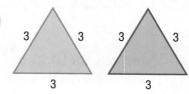

The side lengths of the red figure are:

_____.

The side lengths of the blue figure are:

_____.

The figures have sides that are

_____.

The figures are _____.

Decide if the figures are congruent. Write *congruent* or *not congruent*.

6

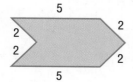

_____

7

_____

Decide if the figures are congruent. Write *congruent* or *not congruent*.

8

_____

9

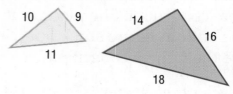

_____

10

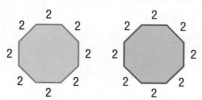

_____

11

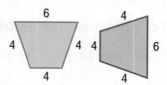

_____

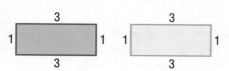

GO ON

Step by Step Problem-Solving Practice

Solve.

Problem-Solving Strategies
- ☑ Draw a diagram.
- ☐ Look for a pattern.
- ☐ Use logical reasoning.
- ☐ Solve a simpler problem.
- ☐ Work backward.

12 **GEOMETRY** Alec wants to know if two figures in his geometry book are congruent. How could Alec use tracing paper and a pencil to find out if the two figures are congruent?

Understand   Read the problem. Write what you know. Alec wants to know if

_____

_____.

Plan   Pick a strategy. One strategy is to draw a diagram.

Solve   Alec placed the tracing paper on top of one figure. What could he do next?

_____

What should he do next?

_____

_____

Check   Model the problem using tracing paper to see if Alec's plan will work.

13 **CRAFTS** Consuela wanted to know if two paper rectangles were congruent. She placed the figures on top of each other. The bottom figure stuck out 2 inches below the top figure. Are the figures congruent?

Check off each step.

_____ **Understand: I circled key words.**

_____ **Plan: To solve the problem, I will** _____.

_____ **Solve: The answer is** _____.

_____ **Check: I checked my answer by** _____.

14 HOBBIES Tokala had two wooden triangles. The figures looked almost exactly the same, but one was a little bit bigger than the other. Tokala decided that the two figures were congruent because they were the same shape. Was he correct? Explain.

_____

_____

15 Reflect How can you be sure that these two rectangles are congruent?

_____

_____

_____

▶ Skills, Concepts, and Problem Solving

Decide if the figures are congruent.

16

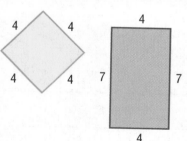

17

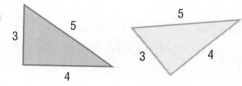

_____

18

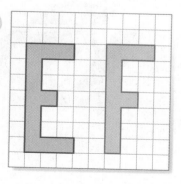

19

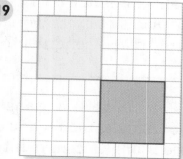

_____   _____

GO ON ▶

Decide if the figures are congruent.

20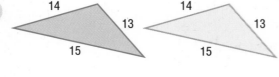

_____

21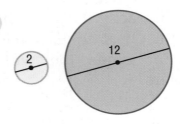

_____

22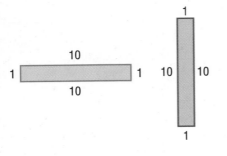

_____

23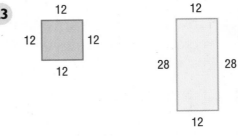

_____

Vocabulary Check **Write the vocabulary word that completes the sentence.**

24 Figures that are the same size and the same shape are called

_____.

25 **Writing in Math** If you carefully trace a figure, will your tracing be congruent with the original figure? Explain.

_____

_____

_____

26 **PICNIC** Rosalita bought a package of paper plates for a picnic. All the plates fit exactly on top of one another in a stack. Are all the plates congruent? Explain.

_____

27 **COMPUTERS** Sunil drew a triangle using his computer. He copied the triangle, then made his second triangle bigger. Are the triangles congruent? Why or why not?

_____

_____

STOP

Reflections

KEY Concept

A **reflection** is one type of **transformation**. A reflection is similar to "flipping" a figure. The **line of reflection** is the line across which the figure is flipped. A reflection does not change the figure's shape or size.

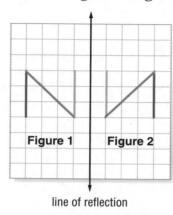

Figure 1   Figure 2

line of reflection

Figure 2 is a reflection of Figure 1.

VOCABULARY

line of reflection
line across which the image or figure is flipped

reflection
a figure that is flipped over a line to create a mirror image of the figure

transformation
a movement of a figure that does not change the size or the shape of the figure

Example 1

Decide if the transformation is a reflection.

1. Check the yellow segments.
   Are they the same distance from the line of reflection? **yes**

   Check the green segments.
   Are they the same distance from the line of reflection? **no**

   Check the blue segments.
   Are they the same distance from the line of reflection? **no**

   Check the red segments.
   Are they the same distance from the line of reflection? **no**

2. If this is a reflection, then all of the corresponding segments will be the same distance from the line of reflection.
   Is the transformation of the figure a reflection? **no**

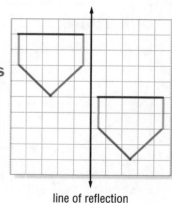

line of reflection

GO ON

Decide if the change or movement is a reflection.

1. Are the **red** segments the same distance from the
   line of reflection? _____

2. Is the movement of the figure a reflection? _____

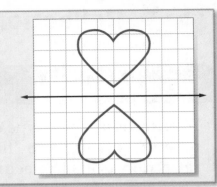

Example 2

Decide which figure is a reflection of Figure A.

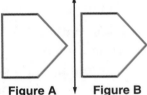  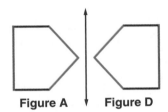

1. Are the **red** segments the same distance from the line of reflection?
   Figure B: **no**   Figure C: **no**   Figure D: **yes**

2. Are the **blue** segments the same distance from the line of reflection?
   Figure B: **no**   Figure C: **no**   Figure D: **yes**

3. Are the **green** segments the same distance from the line of reflection?
   Figure B: **no**   Figure C: **no**   Figure D: **yes**

Figure D is a reflection of Figure A.

YOUR TURN!

Decide which figure is a reflection of Figure A. Write *yes* or *no*.

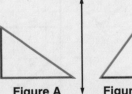

1. Are the **red** segments the same distance from the line of reflection?
   Figure B: _____   Figure C: _____   Figure D: _____

2. Are the **blue** segments the same distance from the line of reflection?
   Figure B: _____   Figure C: _____   Figure D: _____

3. Are the **green** segments the same distance from the line of reflection?
   Figure B: _____   Figure C: _____   Figure D: _____

Figure _____ is a reflection of Figure A.

Who is Correct?

Draw a reflection of Figure 1.

Figure 1

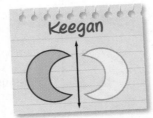

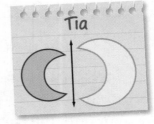

Circle correct answer(s). Cross out incorrect answer(s).

▶ Guided Practice

Decide if the transformation is a reflection.

1 _____

2 _____

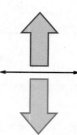

3 _____

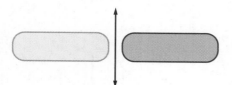

4 _____

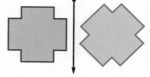

GO ON

5 Determine if the figure is a reflection of Figure A. Write *yes* or *no*.

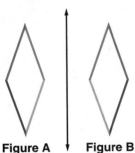
Figure A   **Figure B**

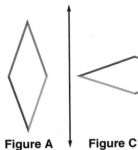
Figure A   **Figure C**

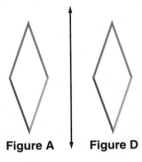
Figure A   **Figure D**

Step 1   Are the red segments the same distance from the line of reflection?

Figure B: _____   Figure C: _____   Figure D: _____

Step 2   Are the blue segments the same distance from the line of reflection?

Figure B: _____   Figure C: _____   Figure D: _____

Step 3   Are the green segments the same distance from the line of reflection?

Figure B: _____   Figure C: _____   Figure D: _____

Step 4   Are the yellow segments the same distance from the line of reflection?

Figure B: _____   Figure C: _____   Figure D: _____

Figure _____ is a reflection of Figure A.

Determine if the figures show a reflection. Write *yes* or *no*.

6

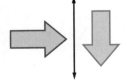   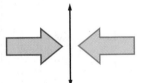   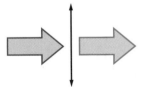

_____   _____   _____

7

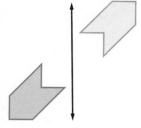   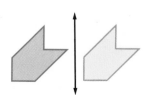   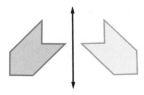

_____   _____   _____

Step by Step Problem-Solving Practice

Solve.

Problem-Solving Strategies
☑ Draw a diagram.
☐ Look for a pattern.
☐ Use logical reasoning.
☐ Solve a simpler problem.
☐ Work backward.

8 DRAWING Kurt was planning to rearrange two posters on one wall of his bedroom. The posters are the same size. He wants to place them evenly spaced on the wall. How can Kurt use reflection to place the posters evenly?

Understand Read the problem. Write what you know.
Kurt has _____ same-sized posters.

Plan Pick a strategy. One strategy is to draw a diagram.

Solve Draw a line of reflection in the middle of the wall.

Then draw one poster on one side of the line of reflection.

Draw the reflection of the poster on the other side of the line of reflection.

Check Are the sides of both posters the same distance from the line of reflection?

9 DRAWING Danielle drew a reflection of a rhombus. Does her drawing show a reflection of a rhombus?

Check off each step.

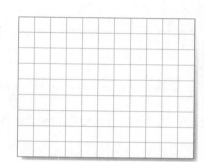

Rhombus A   Rhombus B

_____ Understand: I circled key words.

_____ Plan: To solve the problem, I will _____.

_____ Solve: The answer is _____.

_____ Check: I checked my answer by _____

_____.

GO ON

10 **ART** Lance was reading a book about stained-glass art. A photograph showed two triangle figures that looked exactly alike. They were mirror images of each other. Were they an example of a reflection? Explain.

_____

11 [Reflect] The physical motion of a reflection is called "flipping." Give an example of flipping that you have seen.

_____

_____

▶ Skills, Concepts, and Problem Solving

Decide if the transformation is a reflection.

12
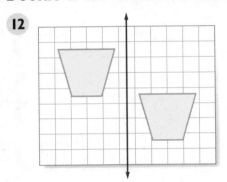

13

14
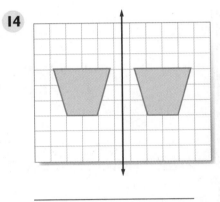

_____

15 Circle the figure that could be a reflection of Figure A.

Figure A   **Figure B**   **Figure C**   **Figure D**

16 Circle the figure that could be a reflection of Figure E.

Figure E   **Figure F**   **Figure G**   **Figure H**

17 **MIRROR** Ofelia was looking at herself in the mirror. Is this similar to what she has learned about reflections in math class? Explain.

_____

_____

18 **DIAGONAL** Tomás noticed that his line of reflection was diagonal. Is he performing a reflection if he flips his figure across a diagonal line of reflection? Explain.

_____

_____

Vocabulary Check **Write the vocabulary word that completes each sentence.**

19 A _____ is a figure that is flipped over a line to create a mirror image of the figure.

20 The line across which the image or figure is flipped is called the _____.

21 **Writing in Math** Some shapes look exactly the same when flipped or reflected. Can you think of two shapes that look the same when reflected?

_____

_____

 **Spiral Review**

Decide if the figures are congruent. Write _congruent_ or _not congruent_. (Lesson 10-1 p. 398)

22 _____

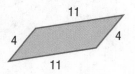

23 _____

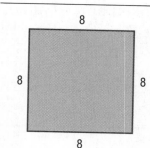

Decide if the figures are congruent. Write yes or no.

1 _____

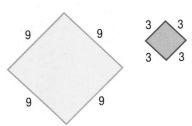

2 _____

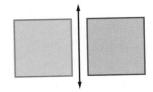

Determine if the figures show a reflection.

3 _____

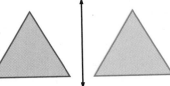

4 _____

Solve.

5 **DRAWING** Does Owen's drawing show a reflection of a triangle? Explain.

_____

_____

6 **PHOTOGRAPH** Lakita was reading a book about modern art. In one photograph, she saw two curved figures that looked almost alike. One figure was a little bit bigger than the other. Was Lakita looking at a reflection? Explain.

_____

Symmetry

KEY Concept

A figure has **line symmetry** when two halves of the figure match.

A figure can have more than one line of symmetry. The line that divides a figure in half so that both parts match is called the **line of symmetry**.

Some figures do not have line symmetry. Some figures have more than one line of symmetry.

This is a line of symmetry. The two halves match.

This figure has two lines of symmetry.

This is not a line of symmetry. The two halves do not match.

VOCABULARY

line of symmetry
a line that divides a figure into two halves that are reflections of each other

line symmetry
a figure has line symmetry when two halves of the figure match

You can check for a line of symmetry. Cut out a figure from paper and fold the paper at the line of symmetry. If the two halves match, then you have found a line of symmetry.

Example 1

Do the figures have line symmetry?

1. Do both halves match? **no**

2. Is this a line of symmetry? **no**

1. Do both halves match? **yes**

2. Is this a line of symmetry? **yes**

GO ON

YOUR TURN!

Do the figures have line symmetry?

1. Do both halves match? _____

2. Is this a line of symmetry? _____

1. Do both halves match? _____

2. Is this a line of symmetry? _____

Example 2

How many lines of symmetry does the figure have?

1. Can the figure be folded vertically to form a line of symmetry? **yes**

2. Can the figure be folded horizontally to form a line of symmetry? **no**

3. Can the figure be folded diagonally to form a line of symmetry? **yes**

4. The figure has 4 lines of symmetry.

YOUR TURN!

How many lines of symmetry does the figure have?

1. Can the figure be folded vertically to form a line of symmetry? _____

2. Can the figure be folded horizontally to form a line of symmetry? _____

3. Can the figure be folded diagonally to form a line of symmetry? _____

4. The figure has _____.

Who is Correct?

Draw all lines of symmetry.

Libby

Luther

Marcel

Circle correct answer(s). Cross out incorrect answer(s).

▶ Guided Practice

Does the figure have line symmetry? Write *yes* or *no*.

1 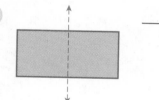 _____

2 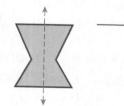 _____

3  _____

4  _____

Step by Step Practice

5 Draw the lines of symmetry.

Step 1 Is there a vertical line of symmetry? _____

Step 2 Is there a horizontal line of symmetry? _____

Step 3 Are there diagonal lines of symmetry? _____

Step 4 Draw all lines of symmetry.

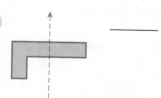

The star has _____ lines of symmetry.

GO ON

Draw the lines of symmetry on each figure. Place an X through the figure if it does not have a line of symmetry.

6

7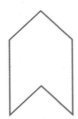

8 L

9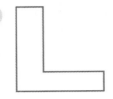

Step by Step Problem-Solving Practice

Problem-Solving Strategies

☑ Draw a diagram.
☐ Look for a pattern.
☐ Guess and check.
☐ Solve a simpler problem.
☐ Work backward.

Solve.

10 **INTERIOR DESIGN** Zina wants to make a sign for her bedroom door with the letters of her name. The letters that have one or more lines of symmetry will be cut out of purple paper. The letters with no lines of symmetry will be cut out of pink paper. What color paper will each letter be on?

Understand   Read the problem. Write what you know. Zina needs to find the _____ on the letters in her name.

The letters with one or more lines of symmetry will be cut out of _____ paper. The letters with no lines of symmetry will be cut out of _____ paper.

Plan   Pick a strategy. One strategy is to draw a diagram.

Solve   Draw the lines of symmetry on the four letters.

ZINA

Which letters have one or more lines of symmetry? _____
Which letters have no lines of symmetry? _____
The letters _____ and _____ will be cut out of purple paper.
The letters _____ and _____ will be cut out of pink paper.

Check   You can check your answer by tracing, cutting, and folding the letters to check for symmetry.

Copyright © by The McGraw-Hill Companies, Inc.

11 DRAWING   Patrick drew a straight line down the exact middle of the pentagon, dividing it into two identical halves. He said he drew the only line of symmetry. Is he correct?

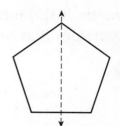

Check off each step.

_____ Understand: I circled key words.

_____ Plan: To solve the problem, I will _____.

_____ Solve: The answer is _____.

_____ Check: I checked my answer by _____.

12 INSECTS   Mindy suggests that the body of a dragonfly has two lines of symmetry. Is she correct? Draw the line(s) of symmetry on the dragonfly.

_____

_____

13 Reflect   Jonathon states that a circle has only one line of symmetry. Is he correct or incorrect? Explain.

_____

_____

 Skills, Concepts, and Problem Solving

Does the figure have line symmetry? Write yes or no.

14  _____   **15**  _____   **16**  _____

Draw all lines of symmetry.

17    **18**   **19**

GO ON

Decide if the figures have a horizontal line of symmetry, a vertical line of symmetry, or both.

20

_____

21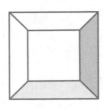

_____

22

_____

Vocabulary Check **Write the vocabulary word that completes each sentence.**

23 A figure has _____ when two halves of the figure match.

24 A _____ is a line that divides a figure into two halves that are reflections of each other.

25 **Writing in Math** Give two examples of everyday objects that have lines of symmetry. How can you check for symmetry?

_____

_____

▶ **Spiral Review**

Decide if the figures are congruent. (Lesson 10-1, p.382)

26

12   4

_____

27

3   3
8   8

_____

Decide if each pair shows a reflection. (Lesson 10-2, p. 389)

28

_____

29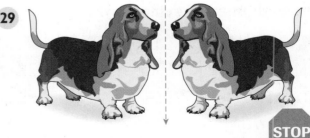

STOP

_____

Lesson 10-4 Translations

KEY Concept

A **translation** is a type of **transformation**.

A translation slides a figure to a new location in a particular direction. The motion of a translation can be vertical, horizontal, or diagonal.

Vertical Translation   Horizontal Translation   Diagonal Translation

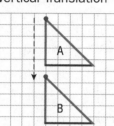   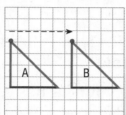   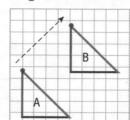

VOCABULARY

transformation
a movement of a figure that does not change the size or the shape of the figure

translation
to move or slide a figure into another position

When a figure changes sizes, it is not a translation. Reflections and turns are not examples of translations.

Example 1

Describe the translation.

1. Are the figures the same shape? **yes**

2. Are the figures the same size? **yes**

3. In which direction(s) did the figure move? **right**

This is a horizontal translation.

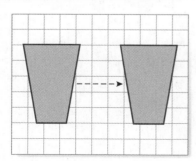

YOUR TURN!

Describe the translation.

1. Are the figures the same shape? _____

2. Are the figures the same size? _____

3. In which direction(s) did the figure move? _____

This is a _____.

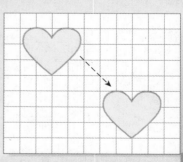

GO ON

Example 2

Show a translation of the figure on the graph.

1. What figure is shown? **It looks like a capital T.**

2. How many squares on the graph are covered? **6 squares**

3. Draw a congruent figure to the right of the given figure.

4. Indicate your translation with an arrow.

5. Do your figures cover the same number of squares on the graph? **yes**

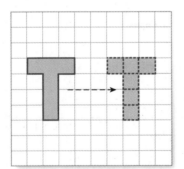

YOUR TURN!

Show a translation of the figure on the graph.

1. What figure is shown?
_____

2. How many squares on the graph are covered? _____

3. Draw a congruent figure to the right of the given figure.

4. Indicate your translation with an arrow.

5. Do your figures cover the same number of squares on the graph?
_____

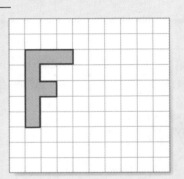

Who is Correct?

Show a translation of the figure below.

Circle correct answer(s). Cross out incorrect answer(s).

▶ Guided Practice

Complete.

1

Are the figures the same size and same shape? _____

Is this a translation? Explain.

_____

_____

2

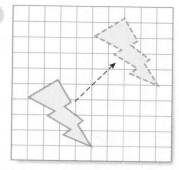

Are the figures the same size and same shape? _____

Is this a translation? Explain.

_____

_____

Step by Step Practice

3 **Describe the translation.**

Step 1 Are the figures the same? _____

Step 2 Are the figures the same size? _____

Step 3 Does this show a translation? _____

Step 4 This is a _____.

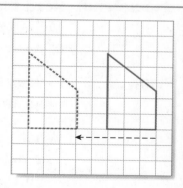

Describe the translations.

4

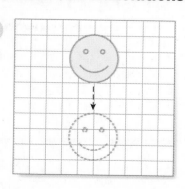

_____

5

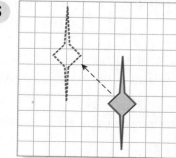

_____

GO ON

Draw a translation of the shape on the grid.

6 vertical

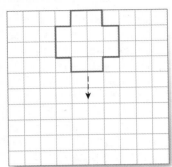

7 diagonal

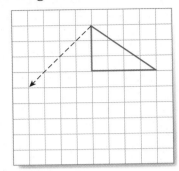

Step by Step Problem-Solving Practice

Solve.

8 **ART** Meliah is decorating a picture frame in art class. She places a diamond-shaped tile on a picture frame. Her art teacher asks her to slide the diamond horizontally. Show Meliah how to perform this translation on the graph paper.

Understand   Read the problem. Write what you know.
Meliah needs to perform a _____
of a _____.

Plan   Pick a strategy. One strategy is to draw a diagram.

Solve   What figure is shown? _____
The diamond is _____ squares across from right to left.
The diamond is _____ squares long from top to bottom.
To translate the diamond horizontally, draw a congruent figure to the _____ of the given figure.

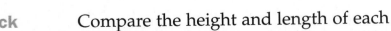

Check   Compare the height and length of each diamond to make sure they are exactly the same.

9 **QUILTING** Sherita placed a star on one of her quilt squares. Then she placed the exact same star four squares to the right. Did Sherita perform a translation? Explain. Check off each step.

_____ **Understand: I circled key words.**

_____ **Plan: To solve the problem, I will** _____.

_____ **Solve: The answer is** _____

_____.

_____ **Check: I checked my answer by** _____.

10 **CALENDAR** Kiyo placed a square around Tuesday to remind him of his soccer practice. Then he drew a circle four spaces to the right around Saturday to remind him of his sister's birthday. Did Kiyo perform a translation? Explain.

_____

_____

11 **Reflect** Why does graph paper make it easier to show an accurate translation?

_____

_____

Skills, Concepts, and Problem Solving

Describe the translation. Write _horizontal_, _vertical_, or _diagonal_.

12

_____

13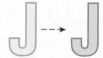

_____

14

_____

15

_____

GO ON

Describe each translation. Write *horizontal*, *vertical*, or *diagonal*.

16 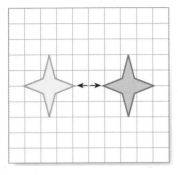 _____

17 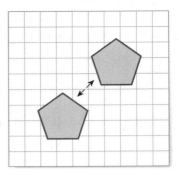 _____

18 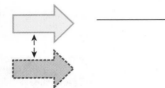 _____

19 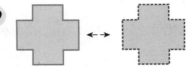 _____

Draw and describe a translation of the figure on the grid.

20 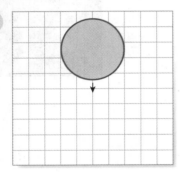 _____

21 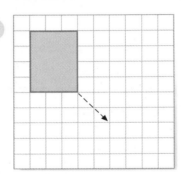 _____

22 DANCING   In Tico's dance class he is learning a new dance routine. He moves seven steps to the left. Is his body performing a translation? Explain.

_____

_____

23 CHECKERS   Marley is playing checkers. She moves her checker piece one diagonal space to the next black square. Did Marley's checker piece perform a translation? Explain.

_____

_____

Vocabulary Check **Write the vocabulary word that completes each sentence.**

24 To move or slide a figure to another position is a _____.

25 A(n) _____ is a movement of a figure that does not change the size or the shape of the figure.

26 **Writing in Math** Imagine that your best friend was absent from school on the day you learned about translations. Explain to your friend how to perform a translation.

_____

_____

 Spiral Review

Decide if the figures are congruent. Write *congruent* **or** *not congruent* (Lesson 10-1, p. 382)

27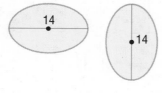

28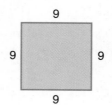

_____

_____

Determine if the figures show a reflection. (Lesson 10-2, p. 389)

29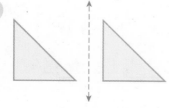

30

_____

_____

Solve (Lesson 10-3, p. 397)

31 **BUTTERFLIES** Kira suggests that the body of a butterfly has one line of symmetry. Is she correct? Draw the line(s) of symmetry on the butterfly.

_____

Draw all lines of symmetry.

1

2

3

Do the figures show a translation? If so, write *horizontal*, *vertical*, **or** *diagonal*.

4
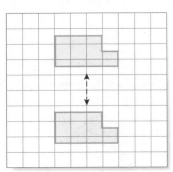

5
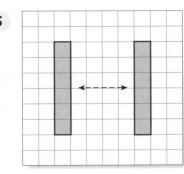

_____   _____

Solve.

6 TRAPEZOID   Isaiah cut out a trapezoid and placed it on graph paper. Then he moved the trapezoid five graph squares below where he first placed the figure. Did Isaiah perform a translation? Why or why not?

_____

_____

7 ART   Angelina made a drawing of a yellow and blue smiling face. She put both drawings at the top of her desk. She moved the blue smiling face under the yellow one. Did Angelina perform a translation? Why or why not?

_____

_____

Vocabulary and Concept Check

congruent figures,
p. 382

reflection, *p. 389*

transformation, *p. 389*

line of reflection, *p. 389*

line of symmetry, *p. 397*

line symmetry, *p. 397*

translation, *p. 403*

Write the vocabulary word that completes each sentence.

1 A _____ is a way a figure can be moved.

2 The line that a figure is flipped or reflected is the _____.

3 _____ is the motion of sliding a figure.

4 Figures that have the same size and same shape are _____.

5 A line that divides a figure into halves that are reflections of each other is a _____.

Write the correct vocabulary term in the blank.

6

This type of transformation is a _____.

7

This type of transformation is a _____.

Lesson Review

10-1 Congruent Figures (pp. 382–388)

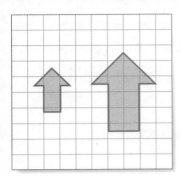

8 Decide if the figures are congruent. _____

Example 1

Decide if the figures are congruent.

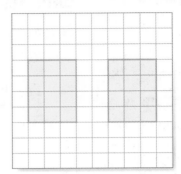

1. Compare the two figures.

2. Do the figures fill in the same number of squares on the grid? **yes**

3. Are they the same shape? **yes**

4. Are they congruent? **yes**

10-2 Reflections (pp. 389–395)

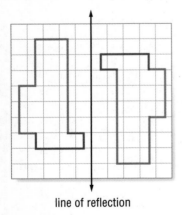

line of reflection

9 Does the diagram show a *reflection*? _____

Example 2

Does the diagram show a *reflection*?

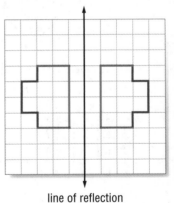

line of reflection

1. Check the red segments. Are they both the same distance from the line of reflection? **yes**

2. Check the green segments. Are they both the same distance from the line of reflection? **yes**

3. Check the blue segments. Are they both the same distance from the line of reflection? **yes**
Does the diagram show a reflection?
yes

10-3 Symmetry (pp. 397–402)

Tell if the line shows a line of symmetry.

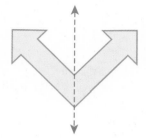

10 Do both halves match? _____

11 Is this a line of symmetry? _____

Example 3

Tell if the line shows a line of symmetry.

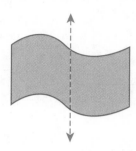

1. Do both halves match? **no**

2. Is this a line of symmetry? **no**

10-4 Translations (pp. 403–409)

12 Decide if the transformation is a vertical, horizontal, or diagonal translation.

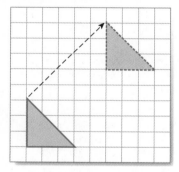

13

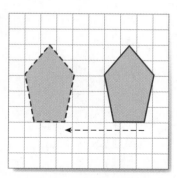

Example 4

Decide if the transformation is a vertical, horizontal, or diagonal translation.

1. Study the figure below.

2. Are the figures the same size? **yes**

3. Are the figures the same shape? **yes**

4. Which direction did the figure move? **up**

The transformation was a vertical translation.

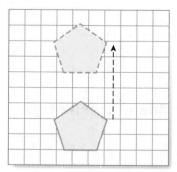

Chapter Test

Decide if the figures are congruent.

1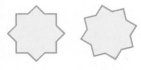

_____

2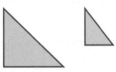

_____

3

_____

4

_____

Circle the figure that could be a reflection of figure A.

5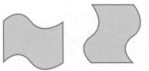

Figure A   Figure B   Figure C   Figure D

Decide if the translation is vertical, horizontal, or diagonal.

6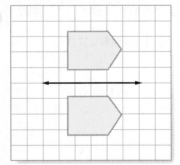

_____

Is the line a line of symmetry?

7

_____

8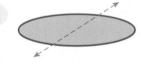

_____

9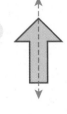

GO ON

Write your answer to each word problem.

10 PUZZLES Graham was comparing two puzzle pieces. The pieces looked almost exactly the same. When Graham held the pieces together, corners of one piece stuck out. Graham decided that the two puzzle pieces were congruent because they were the same shape. Was she correct? Why or why not?

_____

_____

11 ALPHABET Fiona's art teacher told students to find a letter of the alphabet that has a horizontal and vertical line of symmetry. Fiona chose the capital letter H. Does the capital letter H have both lines of symmetry?_____

12 ART Lacey traced a picture of a cat. Then she moved her tracing paper a few inches to the left, and traced the same cat again. Did Lacey perform a translation, a reflection, or neither? Explain.

_____

_____

_____

Correct the mistakes.

13 DRAWING Paul was trying to draw a reflection of a right triangle. Look at Paul's drawing. What did Paul do wrong?

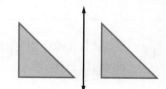

_____

_____

14 STAR Nadia made a drawing of a star. She wanted to perform a translation, so she drew another star four spaces to the right. Nadia made the second star a little bit different from the first. What did Nadia do wrong?

_____

_____

STOP

Choose the best answer and fill in the corresponding circle on the sheet at right.

1 When two figures are the same shape and same size they are _____?

   A reflections

   B lines of symmetry

   C translations

   D congruent figures

2 Which best describes the line of symmetry in the figure below?

   A horizontal line of symmetry

   B vertical line of symmetry

   C both vertical and horizontal

   D no line of symmetry

3 Which two figures show a translation?

4 Jase traced a heart. He moved the paper a few inches to the left and traced the same heart. What transformation did Jase perform?

   A reflection

   B translation

   C line of symmetry

   D none of the above

5 The two triangles below _____.

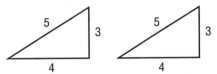

   A are congruent figures

   B have line symmetry

   C are reflections

   D none of the above

6 Which type of translation was performed on the grid below?

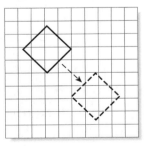

   A horizontal translation

   B vertical translation

   C diagonal translation

   D not a translation

GO ON

7 Which of the figures below have only one line of symmetry?

A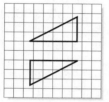

C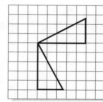

B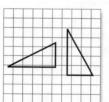

D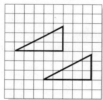

8 Which diagram shows only a translation of the figure?

A

C

B

D

9 Which transformation is shown?

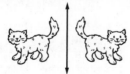

A translation

C reflection

B rotation

D none of the above

10 Tobias is looking at himself in the mirror. Which transformation is Tobias performing?

A reflection

C translation

B rotation

D none of the above

11 Vivian drew a picture of an apple. Beside the apple, she drew a pear. Name the transformation.

A translation

C reflection

B rotation

D none of the above

ANSWER SHEET

Directions: Fill in the circle of each correct answer.

1 (A) (B) (C) (D)
2 (A) (B) (C) (D)
3 (A) (B) (C) (D)
4 (A) (B) (C) (D)
5 (A) (B) (C) (D)
6 (A) (B) (C) (D)
7 (A) (B) (C) (D)
8 (A) (B) (C) (D)
9 (A) (B) (C) (D)
10 (A) (B) (C) (D)
11 (A) (B) (C) (D)

Success Strategy

If you do not know the answer to a question, go on to the next question. Come back to the problem, if you have time. You might find another question later in the test that will help you figure out the skipped problem.

Index